# POPULATION
# AND SOCIETY

In memory of Bob Woods, an imaginative and inspirational population geographer

# POPULATION
# AND SOCIETY

## CLARE HOLDSWORTH, NISSA FINNEY, ALAN MARSHALL AND PAUL NORMAN

Los Angeles | London | New Delhi
Singapore | Washington DC

Los Angeles | London | New Delhi
Singapore | Washington DC

SAGE Publications Ltd
1 Oliver's Yard
55 City Road
London EC1Y 1SP

SAGE Publications Inc.
2455 Teller Road
Thousand Oaks, California 91320

SAGE Publications India Pvt Ltd
B 1/I 1 Mohan Cooperative Industrial Area
Mathura Road
New Delhi 110 044

SAGE Publications Asia-Pacific Pte Ltd
3 Church Street
#10-04 Samsung Hub
Singapore 049483

Editor: Robert Rojek
Editorial assistant: Alana Clogan
Production editor: Katherine Haw
Copyeditor: Rosemary Morlin
Indexer: Rachel Gee
Marketing manager: Michael Ainsley
Cover design: Wendy Scott
Typeset by: C&M Digitals (P) Ltd, Chennai, India
Printed by MPG Printgroup, UK

**Library of Congress Control Number: 2012944705**

**British Library Cataloguing in Publication data**

A catalogue record for this book is available from
the British Library

MIX
Paper from
responsible sources
FSC
www.fsc.org    FSC® C018575

ISBN 978-1-4129-0064-5
ISBN 978-1-4129-0065-2 (pbk)

# CONTENTS

# ABOUT THE AUTHORS

Clare Holdsworth is Professor of Social Geography at Keele University. She trained originally as a demographer and her research interests have covered many aspects of population geography, including historical demography and contemporary household and family formation. Her most recent book *Family and Intimate Mobilities* (Palgrave Macmillan, 2013) considers the interrelationships between family practices and mobility.

Nissa Finney is a Hallsworth Fellow at The University of Manchester, based in the Cathie Marsh Centre for Census and Survey Research. Nissa's training is in Geography and her research and teaching focus on migration, ethnic inequality and mixed methods of research. She is the author (with L. Simpson) of '*Sleepwalking to Segregation'? Challenging Myths of Race and Migration* (Policy Press, 2009).

Alan Marshall is a Research Associate based at the Cathie Marsh Centre for Census and Survey Research within the University of Manchester. He is a social statistician by training and has research interests in health inequalities and population ageing.

Paul Norman is a Lecturer at the School of Geography, University of Leeds. He is a population geographer/demographer with an interest in how area type and population change, in particular migration, influence health inequalities. Paul's on-going research investigates demographic and health change by ethnic group.

# ACKNOWLEDGEMENTS

This book was first conceived as a project in the Department of Geography, University of Liverpool. Clare Holdsworth would like to express her thanks to Bob Woods, who sadly passed away before the book was completed and who very generously provided material for the manuscript which is reproduced in chapters 3 and 5. She would also like to thank Paul Williamson and Bill Gould for their enthusiasm for the project and for encouraging her to continue with the book and her co-authors who so readily rose to the challenge of contributing to the book and shaping its form and direction.

The authors are grateful to students at Keele, Liverpool, Manchester and Leeds universities who have engaged with many of the ideas in the book and provided a sounding board for how to teach population geography. The authors have common ground in that they have all worked at the Cathie Marsh Centre for Census and Survey Research (CCSR) though rarely at the same time! They would like to thank colleagues at CCSR for providing such a stimulating environment to research population geography, particularly Angela Dale and Ludi Simpson for their invaluable mentorship and support.

We would like to thank Robert Rojek at Sage for his continual support for the project and Alana Clogan for guiding us through finalising the manuscript. We are especially indebted to Andrew Lawrence at Kudis in Keele University for drawing all the figures.

The authors are grateful to the following for permission to use copyright material:

United Nations Children Fund (2002) *The State of the World's Children 2002: Leadership.* Geneva: UNICEF.

United Nations Children Fund (2008) *The State of the World's Children 2008: Child Survival.* Geneva: UNICEF.

United Nations Department of Economic and Social Affairs Population Division (2000) *Replacement Migration: Is It a Solution to Declining and Ageing Populations?* New York: United Nations. Available at http://www.un.org/esa/population/publications/migration/migration.htm. Accessed December 2011.

United Nations Department of Economic and Social Affairs Population Division (2006) *Trends in Total Migration Stock: The 2005 Revision.* New York: United Nations. Available at: http://www.un.org/esa/population/publications/migration/UN_Migrant_Stock_Documentation _2005.pdf

United Nations Department of Economic and Social Affairs Population Division (2009) *Trends in International Migrant Stock: The 2008 Revision.* New York: United Nations. Available at: http://www.un.org/esa/population/migration/UN_MigStock_2008.pdf. Accessed April 2011.

United Nations Department of Economic and Social Affairs Population Division (2009) *World Population Prospects. The 2008 Revision.* New York: United Nations. Available at: http://www.un.org/esa/population/publications/wpp2008/wpp2008_highlights.pdf. Accessed December 2011.

United Nations Department of Economic and Social Affairs Population Division (2011) *World Population Projections 2010 Revision.* New York: United Nations. Available at: http://esa.un.org/unpd/wpp/unpp/panel_population.htm. Accessed December 2011.

United Nations Department of Economic and Social Affairs, Statistics Division (2005) *Demographic Yearbook 2004.* New York: United Nations. Available at: http://unstats.un.org/unsd/demographic/products/dyb/dyb2.htm. Accessed June 2010.

United Nations Department of Economic and Social Affairs, Statistics Division (2006) *Demographic Yearbook 2005.* New Year: United Nations. Available at http://unstats.un.org/unsd/demographic/products/dyb/dyb2.htm. Accessed June 2010.

United Nations Department of Economic and Social Affairs, Statistics Division (2008) *Gender Info 2007.* New Year: United Nations. Available at http://unstats.un.org/unsd/demographic/products/genderinfo/. Accessed June 2010.

OECD (2011) Family Database, OECD, Pairs, Retrieved December 2011 from www.oecd.org/social /family/database

OECD (2007) Social Expenditure database 1980–2003. Retrieved December 2009 from www.oecd.org/els/social/expenditure

Taylor and Francis for permission to reproduce material from Preston, S. (1975) 'The changing relation between mortality and level of economic development', *Population Studies* 29(2) 231–248.

John Wiley and Sons for permission to reproduce material from Reher, D. (2004) 'The demographic transition revisited as a global process', *Population Space and Place* 10: 19–41.

Hafan Books for permission to reproduce material from Charles, E.N., Cheesman, T. and Hoffmann, S. (eds) (2003) *Between a Mountain and a Sea. Refugees writing in Wales.* Swansea: Hafan Books.

US Bureau of the Census for permission to reproduce material from Smith, K., Downes, B. and O'Connell, M. (2001) 'Maternity leave and employment patterns: 1961–1995', *Current Population Reports*, pp. 70–79. Washington, DC: United States Census Bureau.

UK Census output is Crown copyright and is reproduced with the permission of the Controller of HMSO and the Queen's Printer for Scotland.

# 1

# INTRODUCTION

## Key population questions

At the time of writing this book, October 2011, the global population was expected to surpass 7 billion living souls. By the time you are reading this book, the global population will easily have passed this milestone, and maybe the fear or optimism of exceeding this number is now a thing of the past, or maybe not. Current projections of global population estimate that the number of people living on our plant will increase to 9 billion by 2050. In contrast, in 1900, world population was around 1.6 billion and in the year 1000 was probably around 310 million. There is no doubt that the world has experienced a period of unprecedented population growth. World population has doubled in 38 years, and is currently increasing at the rate of 78 million people per year. Yet while these headline figures of world population numbers and growth are widely publicised, it is more challenging to state precisely what the consequences of this change will be. This can be taken as a basic tenet of population studies: to understand, explain and predict the causes and consequences of population change. There are a number of different ways of tacking these issues, some of which have attracted much controversy over the years. Robert Malthus, Karl Marx, Paul Ehrlich, Ester Boserup, Julian Simon, Amartya Sen and many others have discussed and predicted the critical relationship between population and resources. Successive generations of scientists have struggled to accumulate and refine the empirical evidence of these claims and there are still many points of fact and interpretation which are open to debate. Not least among these are the likely future rates of world population change, its sustainability, and its environmental, economic, social and political consequences.

Thus while headline measures of population growth and key milestones, such as exceeding 7 billion, generate considerable popular and political interest in population, interpretation and validation of these trends and projections is less straightforward. In fact, population estimation and projection is not a new science and a cursory review of early attempts to both measure and project population size and associated consequences illustrates that it is certainly not an exact one. Consider the following:

1　In the late seventeenth century an amateur Dutch scientist, Antoni van Leeuwenhoek, attempted one of the first estimates of global population, when he proposed that the number of people in the world could not be more than 13.4 billion. Van Leeuwenhoek's conclusion, although technically correct, was very inaccurate as historians estimate that the global population in the latter 1600s would have been closer to half a billion (Kunzig, 2011).

2　In the 1930s, official projections of the population of England and Wales suggested that by the year 2000 the total population would be somewhere between 17.5 and 28.5 million; in contrast in the 1960s, population projections predicted the population of England and Wales would surpass 65 million by the beginning of the third millennium and concerns were expressed about the need for urban expansion to accommodate this population (Hobcraft, 1996). In fact the 2001 Census for England and Wales recorded a total population of just over 52 million.

3　In 1980 two population scientists, Julian Simon and Paul Ehrlich, with opposing views on the potential or threats of global population growth had a bet over the future of commodities prices. Ehrlich chose five commodities – copper, chromium, nickel, tin, and tungsten – Simon bet that their prices would decrease between 1980 and 1990. Ehrlich bet they would increase, as global population growth became unsustainable. Simon won the bet.

So how can we explain the disparities between population estimates and projections on the one hand and realised population counts on the other? Why do population scientists disagree about the global outcomes of continued population growth? In part, the answer to the first question is that as population geographers and demographers have studied the dynamics of population growth (and decline) over time, we have learnt more about why population increases, sometimes quite dramatically, while at other times population growth stalls, or even goes into decline. Our methods for collecting reliable data on population, both counts of the number of people living in a particular locality and accurate records of demographic events, namely births and deaths, have improved greatly over the last two centuries. In the seventeenth century, scientists could only speculate about global population; van Leeuwenhoek's calculation of global population was based on an extrapolation of how many people he estimated to be living in the Netherlands and the relationship between the size of the Netherlands and the Earth's habitable land. Now population scientists can count population size more reliably, using census counts, administrative records and sample surveys. Moreover, we have become more accustomed to and accommodating of official apparatus to collect key demographic data.

However, being able to make accurate population projections depends not only on reliable measures of current population size and underlying trends in mortality and fertility (and migration for national or regional projections), but also of a population's recent past. Fears about population decline in England and Wales in the 1930s were driven by low levels of fertility recorded at that time, but these were not sustained after the Second World War, when fertility began to increase again. The 1960s baby boom was brought about in some part by women having their children at relatively younger ages than earlier generations. What the population researchers in England and Wales in the 1960s did not foresee was that by the end of the decade and into the 1970s, women would start to have their children at older ages, thus lowering overall levels of fertility. The population projections in the 1930s and 1960s extrapolated contemporary fertility and mortality rates into the future, yet neither the population projections of the 1930s nor 1960s were realised as the population dynamics experienced at those particular times were not carried forward. It is important to distinguish between population projections, which are based on a continuation of current rates, and population forecasts, which predict what a future population might be.

While the reliability of population projections has improved, and lessons have been learnt from the inaccurate projections of the past, interpreting the consequences of population growth is still very much up for grabs. At the time of writing, as global population passes another important milestone, the popular tone is more pessimistic than optimistic, and a cursory glance at media articles to commemorate a global population of 7 billion would not give much support to Simon winning his bet again if it was wagered in the twenty-first century. A populist reading of current debates about population would suggest that Simon's optimism about the potential of population growth to stimulate economic growth and technological innovation would appear less sustainable as global population continues to rise, though many population researchers continue to share Simon's belief that it is possible to feed both the current and future global population. Yet it is all too easy to worry about headline figures, while ignoring the detail. While our starting point for a text on population and societies is headline figures of population totals and growth, we should stress that these can only be a starting point. The purpose of writing this book is not to argue on the side of either Ehrlich or Simon, or to mull over the future consequences of global population growth, but rather to understand how we have got to where we are and the challenges and opportunities of contemporary population issues. Our premise is that population matters, and it matters if we are concerned about global population of 7 billion and rising; or dealing with population decline in disadvantaged urban neighbourhoods; or with the dynamics of an ageing society; or how to find opportunities for youthful populations in parts of the developing world; or understanding the causes and implications of the HIV-AIDS endemic in Sub-Saharan Africa; or how to tackle regional health inequalities in the UK. Our list of population issues could go on and could include all types of population events at different scales; from individual, to inter-urban to regional, to national to global.

Thus the themes that we develop throughout this book are not directed towards absolute numbers but rather the dynamics of population change. We identify three broad themes: the significance of scale; the magnitude of inequality; and the importance of data and their interpretation. When thinking about scale, the dynamics of population change can alter greatly depending on both the geographical and/or temporal scale being used. Take for example global population growth: while the world's population is predicted to increase, this global increase is brought about by an assemblage of population growth and decline in different countries and subnational regions. Scale is also relevant to our second theme of inequality; the demographic unevenness identified at different times and places has consequences for how individuals access resources and how these resources are distributed. Moreover, solutions to inequalities might vary depending on unit of analysis, comparing for example distribution of resources within and between communities, cities, nations or global regions. Finally, if we are both to describe and explain the dynamics of population growth, we need to have reliable data and accurate ways of measuring the components of demographic change. Moreover, we need to understand how demographic indices are computed and the assumptions that are made in their calculation.

We argue that population matters, but this might raise the question 'for whom?' If we want to consider who takes responsibility for demographic change, then a reasonable answer is that we all do, and that responsibility rests not just with individuals, but also with communities, government and government agencies at all scales from intra-urban districts to national assemblies, and international agencies. Thus as individuals we have responsibility for our health, fertility choices and mobility and migration decisions. Yet these decisions and behaviours are made and take place within specific contexts; a key issue that population researchers have considered is the relationship between individual agency on the one hand and the role of government policy and social structures on the other. Responsibility cannot just lie with individuals, as both political and social institutions have to be able to respond to the dynamics of population change, and, where appropriate, bring about changes in individual behaviour or alleviate the impacts of change.

Since the mid-twentieth century some of the most pressing demographic issues, particularly relating to inequalities between the global North and South, have been considered by the United Nations (UN) and other international non-government organisations (NGOs). The need for international co-operation in meeting the challenges of population change has been a core theme of UN policy and programmes. Between 1954 and 1994, the UN organised a decennial World Population Conference (the inaugural conference was held in Rome in 1954 and the 1994 conference was in Cairo); in 1999 a special session of the UN General Assembly was convened to review and appraise the implementation of the Programme of Action adopted at the 1994 Cairo Conference. In 2000, the UN published its Millennium Development Goals (MDGS). There are eight goals in total (which are listed in Box 1.1), none of which refer to population growth per se, but all have a bearing either directly or indirectly on population change. For our purpose, the significance of the MDGs is that they identify the need for co-operation in responding to the challenge of inequality, the second theme of this book. The goals also reveal the importance of socio-demographic research and data in capturing and monitoring the conditions of populations across time and at different geographical scales.

---

### Box 1.1   UN Millennium Development Goals

1   Eradicate extreme poverty and hunger
2   Achieve universal primary education
3   Promote gender equality and empower women
4   Reduce child mortality rates
5   Improve maternal health
6   Combat HIV/AIDS, malaria, and other diseases
7   Ensure environmental sustainability
8   Develop a global partnership for development

---

## How to use this book

This book is intended to introduce readers to the main themes of population dynamics and population geographies. We do not claim to provide a definitive account of population issues; rather we seek to draw readers' attention to the complexity of key population trends, and to appreciate the interconnectivity of population issues, both in terms of disentangling the key events of births, deaths and migrations, as well as the implications for social, political, cultural and environmental change. Inevitably, for a book on the topic of population, we present a large number of graphs and data. Our use of data and diagrams is not to illustrate the text, but to demonstrate the dynamics of population events. We recommend that readers spend as much time interpreting the diagrams and getting to grips with the data as reading the text. Clearly we do not have the space to illustrate the totality of global population trends, but rather we select data for countries that can illustrate the main global, as well as regional, processes. We do not restrict our account of population dynamics to national comparisons, but also consider population variation at subnational scales. Inevitably population data becomes dated, and there are delays in accessing up-to-date data. In particular, population counts are most readily available from population censuses, which are usually carried out every ten years. At the time of writing in 2011, for example, UK census data for that year are not available. Moreover, if we want to compare comparable data from different countries, this may also restrict data availability. Readers might find it useful to refer to the websites of major NGOs, provided in the resources section of each chapter, which can provide regular updates on demographic indicators.

At times our focus is more towards the developed world and the UK, particularly considering the dynamics of subnational population change. In part, this reflects our own expertise as British demographers and population geographers, but also because there are other texts available that deal very thoroughly with issues linking population and development; and where appropriate we direct readers to these resources. Our case studies are, however, used to illustrate particular processes and trends and do not just pertain to the countries or regions that we use in the book. You might want to consider how the data we use compares with that for the country or region where you are reading this book. At the end of each chapter, we suggest further readings to find out more about the topics and issues covered. Thus this text should be approached as introductory and one that will hopefully stimulate readers to find out more about the complexities and intricacies of the relationship between population and societies.

## Book structure

The text considers different aspects of population in turn. We have tried wherever possible to minimise repetition though we have not avoided this entirely. Our approach is to begin with the bigger picture, in that we start by considering population dynamics in their entirety before considering their constituent parts. We hope by beginning with transitions and structures that this emphasises the need to consider the complexity of demographic change. Population change can never be just about one element; to understand fertility decline we need to consider what has happened to mortality, and to be able to predict future fertility we need to have an understanding of migration, and so on.

The first substantive chapter considers the concept of transition. The logic of transitions has been used by population scientists to understand historical patterns of population growth and stabilisation, and how these are brought about by changes in fertility and mortality. Our review compares different approaches to transitions and in particular the extent to which the demographic transition model can be interpreted as a descriptive or an analytical tool. What is at issue here is unravelling the relationship between long-term mortality and fertility decline and this chapter reviews some of the main explanations of declining vital rates. We also consider how the concept of transitions has been applied to other dimensions of demographic change, specifically health and migration.

The concept of the demographic transition is well established in population studies, and reveals how researchers have approached the dynamics of population change over the last century. In Chapter 3, we turn to consider how population researchers use data on past and current population dynamics to make population projections. In this chapter, we also outline the main data sources and the kinds of considerations and assumptions that we need to be aware of when accessing and using demographic data.

Knowledge of population size and structure is fundamental to both population research and applied work about populations, and in Chapter 4 we investigate the age-sex structure of population and how these vary over time and place.

In Chapter 5, we turn our attention more closely to fertility and mortality and outline the main approaches to measuring and comparing vital rates. We consider fertility and mortality together as there are some commonalities in the calculation of basic rates. While this chapter outlines the main demographic approaches to measuring vital rates, it also considers how the computation and interpretation of vital rates is essential for understanding the underlying causes of fertility and mortality decline and increase and how these vary over time and place.

Chapter 6 considers the third component of population change: migration. This chapter reviews definitions of migration and ways in which migration is measured. Models of migration processes are introduced and several theoretical approaches that aim to understand the causes and consequences of migration are outlined. The chapter provides some national and subnational examples of migration.

Chapter 7 on living arrangements turns to consider the formation and structure of families and households and their diversity over time and place. In doing so, we seek to challenge some commonly held beliefs about families and households in present and past times. We consider the causes and implications of two important trends in living arrangements: the feminisation of households and the increased propensity to live alone.

The theme of family is continued into Chapter 8 on family formation and fertility. In this chapter, we outline how a key theoretical concept of individualisation can help us to understand more recent changes to patterns of leaving home, partnership formation and dissolution as well as fertility, particularly in modern industrialised societies. We also continue with the theme of transition by reviewing how demographers have developed the concept of the second demographic transition to describe and explain how changes in family formation and fertility are interlinked.

Risks of dying and suffering long-term illness are not uniformly distributed and in Chapter 9 we investigate how the chances of an individual living a long and healthy life varies according to a number of factors including their country of residence, their socio-economic circumstances and the particular neighbourhood in which they live. In this chapter, we also outline various theoretical explanations of the causes of health inequalities.

In the final chapter, we turn our attention to population futures and review how two important population issues – overpopulation and population ageing – have been researched and debated by population researchers. We also consider the challenges facing the next generation of population researchers in terms of dealing with inequalities, methods of data collection and access to data and the importance of scale in a globalised world.

# 2

# TRANSITIONS: THINKING ABOUT POPULATION OVER TIME

Populations are constantly changing in both size and structure. At the beginning of the twenty-first century, as we outlined in the previous chapter, a main concern is population growth and the challenges and opportunities that this brings. Yet our understanding of population change is not only about anticipating the future, but also involves population change in past times and unravelling the complex processes and causality of population history.

The dramatic population increases of the late nineteenth and twentieth century were fuelled by changes in the main components of the population equation: mortality, fertility and migration. In this chapter, we consider how these components interact to create the conditions for population change and how demographers have used the concept of transition to represent and explain population change over time. Population scholars use the transitions framework to conceptualise the tension between population and modernity, which is essentially a way of understanding the relationship between population and time. Transitions can also be used as a way of conceptualising the relationship between aggregate populations and individual behaviour. Thus, the idea of transitions incorporates different understandings of time that relate to both aggregate populations and individuals; that is looking at aggregate population transformations over historical periods of chronological time, as well as individual change over the life course. This chapter reviews the different ways in which population geographers and demographers have engaged with and used the transitions framework to understand population change in relationship to population growth and modernity, migration and epidemiology.

Demographic change is not just a matter for national populations, but is relevant and dynamic at subnational scales. Though the conventional concern with population and structure has tended to reify national boundaries as the meaningful identifier for population processes (for example, the tendency to refer to the UK's population, fertility or mortality), this focus on nations can overlook how demographic change is experienced at subnational, regional, community and family/individual scales (as well as at supranational scales). Thus we conclude this chapter by considering how scale is relevant for demographic transitions.

This chapter addresses the following questions and issues:

- What was the relationship between fertility and mortality in pre-transition societies?
- How have demographers developed the concept of the demographic transition to represent national-level changes in population size and growth?
- Is the demographic transition model a descriptive or an explanatory tool for understanding population change?
- What are the main explanations of long-term mortality and fertility decline?
- How has the concept of transition been applied to other aspects of demographic change, in particular causes of mortality and migration?
- How have demographers used the concept of the life course to understand the pattern of events over an individual lifetime and what is the relationship between individual behaviour and aggregate population change?

## Population change

At the simplest level, populations grow because there is a positive balance between births and deaths and this positive balance is not removed by net out migration. For example, consider world population growth for the decade 2000 to 2010. The population in 2010 ($P_{2010}$) will be the population in 2000 ($P_{2000}$) plus the births added in the decade (B) minus the deaths (D). So we can write:

$$P_{2010} = P_{2000} + B - D \qquad (1)$$

It is reasonable to assume that as the world population is a closed system, migration will have no influence, so the case of world population growth is the most straightforward. However, if we considered the population growth in the UK then we would need to use the following equation:

$$P_{2010} = P_{2000} + B - D + I - E \qquad (2)$$

where I stands for the number of immigrants (in-migrants) and E the number of emigrants (out-migrants) during the same period of time. We can take this equation and use it to define a general equation for population growth. Distinguishing any point at time $t$, and a further point at time $t+n$, we can rewrite the above equation:

$$P_{t+n} = P_t + B_{t,t+n} - D_{t,t+n} + I_{t,t+n} - E_{t,t+n} \qquad (3)$$

Equation (3) is the basic demographic equation. It defines the ways in which populations change, that is through natural increase (B–D) and net migration (I–E):

$$\text{Population change} = \text{Natural Increase} + \text{Net Migration} \qquad (4)$$

The causes of population change are highly variable; in some parts of the world at certain historical times migration will have the greatest impact, while in other places and times natural increase will dominate. This equation is often expressed as rates, by dividing each component by the mid-year population. The calculation of crude death and birth rates, the most basic demographic measures of mortality and fertility is outlined in Box 2.1. More precise measures of mortality and fertility are discussed in Chapter 5.

---

### Box 2.1    Crude birth and death rates

Most of the indices used by population scientists are derived as a ratio (or rate) of a numerator (N) to a denominator (D) expressed in parts per 1000:

$$\frac{N}{D} * 1000$$

The numerator gives the number of events occurring for a particular time period (often a year but can be longer), while the denominator represents the vulnerable, or 'at risk' population. For a rate measured over one year, the denominator is measured as the mid-year population.

The crude birth and death rates are the most straightforward and easy to compute measures of fertility and mortality. They are defined as 'crude' as they do not take into account the age structure of the population at risk, as the denominator is the total population:

$$\text{Crude birth rate} = \frac{\text{number of live births in a year}}{\text{Mid-year population}} * 1000$$

$$\text{Crude death rate} = \frac{\text{number of deaths in a year}}{\text{Mid-year population}} * 1000$$

The crude rate of natural increase is derived from the crude birth and death rates. It is simply the difference between the two rates:

Crude Rate of Natural Increase = Crude Birth Rate – Crude Death Rate

Equation (1) provides an important first step to understanding the dynamics of population change. It should now be possible to see how populations might grow or decline over time, to begin to appreciate that the balance between fertility, mortality and migration is likely to vary under different economic, social, political, environmental and epidemiological conditions. Geographical scale is also important as the growth of small populations is likely to be particularly influenced by migration while in large countries and world regions the factors influencing natural population growth will be of greater significance. Figure 2.1 (overleaf) plots the timepaths of CBR and CDR for six world regions using data from the UN *Demographic Year Book* for the six periods: 1953–7, 1960–4, 1970–5, 1980–5, 1990–5 and 2005–10. The diagram has been plotted to illustrate the effects of changes in CBR and CDR (which is plotted on reverse scale on the horizontal axis), with the crude annual rate of population growth on the diagonal line. On the zero diagonal line, CBR and CDR balance one another and there is no population growth, on the 3-per cent line CBR is sufficiently in excess of CDR to generate rapid growth.

Figure 2.1 indicates that world population growth has remained at 1.6 to 1.8 per cent per year since the 1950s, although by now it will be less than 1.5 per cent and falling. But what is most interesting in the graph is the very different relationship between fertility and mortality in the major world regions. Let us begin by considering Europe. Since the 1950s, CBR in Europe had continued to drift downwards, while the CDR had begun to increase as the structure of the population has aged. By the mid-1990s, Europe's population had reached zero growth with CBR and CDR equal at 12 per 1,000. North American is approaching zero growth; American CDR has remained relatively unchanged at nine per 1,000, and the CBR has declined substantially over recent decades, so that the rate of growth is now only 0.5 per cent per year. Turning to the remaining four less developed regions, the growth rate for these regions is now approaching 1 per cent a year following substantial declines in CBR. In South Asia (India, Pakistan, Bangladesh and Sri Lanka) CBR has begun to decline, while the CDR is also falling, so that this population growth is less than 2 per cent per year. Latin American has experienced a rapid decline in fertility since the early 1960s. Whereas growth once appeared to be heading for 3 per cent it is now less than 2 per cent per year and falling. Africa is the latest world region to show signs of decreasing growth rate due to some reduction in fertility, but until the early 1990s growth rates increased because of declines in mortality.

What can we make of the different demographic timepaths in Figure 2.1? We might be tempted to take Europe's attainment of zero growth as denoting the end of a journey, or the end of a growth cycle. Other regions are still travelling on this journey, but they will eventually complete it, albeit at different speeds. We might confidently predict that the world's rate of population growth will eventually approach zero; it will have a stationary population with births and deaths equalling one another. However, predicting when this will occur is another matter altogether. Moreover, the starting point for these demographic journeys is the 1950s, and even for this time period the analysis is based

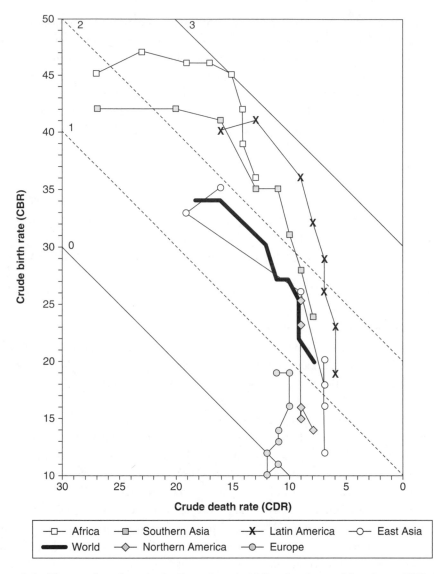

Figure 2.1    Time paths of crude death and crude birth rates for world regions, 1950–2010

*Source:* UN Demographic Yearbooks 1950 to 2010

on estimates of fertility and mortality. However, it would be fascinating to take these timepaths back to an earlier period. The implications of taking a longer historical perspective are discussed in the next section of this chapter on demographic transitions. The fundamental components of population growth do not remain in equilibrium, but at specific times in history the relationship between demographic events shifts. It is these changes that the demographic transition model summarises, and particularly those changes associated with modernity. In the next section, we explore the dynamics of pre-transition populations before considering the process of transition itself.

## Malthus and pre-transition population dynamics

The demographic transition is concerned with the synchronicity between mortality and fertility decline which brings about population growth, then stabilisation of growth. Yet before turning to this model in detail, it is appropriate to consider the dynamics of population in what we might describe as 'pre-transition' societies, that is where both fertility and mortality are high. The most (in)famous contributor to our understanding of the balance between mortality and fertility in these kinds of conditions is the Revd Thomas Malthus, author of *An Essay on the Principle of Population* first published in 1798, with a further five editions published between 1798 and 1826 (see Winch, 1987). Malthus was concerned with the relationship between population growth and resources. In his first essay, Malthus set out his theory of the relationship between population and resources, based on two fixed certainties·

1   food and passion between sexes are both essential for humankind's existence;

2   whereas population grows at a geometric rate, the production capacity only grows arithmetically.

Thus as population growth would quickly exceed that of agricultural production this would necessitate one of two forms of checks on population growth. The first, and the one which Malthus is most popularly associated with, is a positive or unplanned check of war, pestilence and famine. The second check and one which Malthus wrote more about in subsequent editions of his essay, is a preventative or planned check through which couples reduced their fertility; this could be achieved by marrying at older ages and sexual constraint within marriage. Thus as Winch (1987) describes, over the first two decades of the nineteenth century Malthus developed his ideas about population and became more interested in the potential of moral constraint; and put more emphasis on how the logic of his two certainties about population and agricultural growth could be mitigated by individual responsibility.

It is important to consider the political, economic and social context in which Malthus wrote his essays and why a rural provincial clergyman became one of the most celebrated, or infamous, theorists of the perils of population growth. In writing his essays, Malthus was engaging not just in a discussion of the scientific logic of population growth and of the importance of moral constraint, but was also contributing to wider debates that dominated late eighteenth- and early nineteenth- century political life. In particular, Malthus was critical of the English Poor Laws, a system of relief for the poor which had developed in the sixteenth century out of earlier Tudor relief systems. By the end of the eighteenth century, Malthus was certainly not alone in his belief that rather than alleviating poverty, the Poor Laws encouraged pauperism. While Malthus acknowledged that the Poor Laws alleviated 'very severe distress', overall 'the aggregate mass of happiness among the common people would have been much greater than it is at present' if the Poor Laws had not been in place (quoted in Winch, 1987: 43). According to Malthus, the problem with the Poor Laws was that in maintaining a minimum level of support during scarcity they had 'contributed to the problem they were designed to alleviate by lowering wages, increasing the price of food, and encouraging population increase' (Winch, 1987: 43). Malthus was also reacting to other cultural changes, in particular the promotion of sexual liberation and female autonomy associated with the writings of William Godwin and his wife Mary Wollstonecraft (Bederman, 2008). Writers such as Godwin believed in the 'perfectibility of society' and the potential for humankind and were influenced by the tumultuous events of the French Revolution. In contrast to Godwin's optimism about human potential, from Malthus's scientific perspective a limit to human growth was both desirable and inevitable.

We shall return to consider how Malthus's ideas have been developed and disputed by scholars over the last 200 years in the final chapter. The reason for discussing Malthus in a chapter on transitions is that Malthus described a closed system where population growth will ultimately be brought in check by increases in mortality; a relationship between growth, resources and mortality that can be described as a 'Malthusian trap', and the most moral solution to this trap is to reduce the potential for growth. Yet what has fascinated demographers is that the historical record demonstrates that Malthus was wrong about the impossibility of sustained population growth. In the nineteenth century and beginning in Northern Europe, mortality declined bringing about both population growth and ultimately fertility decline. It is this synchronicity between changes in death and birth rates that the Demographic Transition Model considers.

## The demographic transition model (DTM)

The DTM is an enduring concept in the study of population. It proposes that a country's population change will go through a series of stages (in association with other social and economic changes) where the changing balance of births and deaths determines overall population change. The first recognisable variation of the model was developed by the American demographer Warren Thompson in 1929, though Thompson's model was purely descriptive. The first formulation that sought to use the transition approach to *explain* population change, and specifically fertility changes, was developed by Frank Notestein in 1945. Since Notestein's paper, the concept of the transition model has been considered by most of the influential demographers of the twentieth century. Contributions by, for example, Jack Caldwell, John Cleland, Ansley Coale, Tim Dyson, Dudley Kirk and Simon Szreter have sought in various ways to consolidate, develop or challenge the model (for a recent review see Dyson, 2010). One reason for the model's enduring popularity is the simplicity of the idea, which has certainly made the model amenable to teaching introductory population studies. If you have had some exposure to population studies, you will probably have already learnt a variation of the model, mostly as a way of describing the *key* features of historical population change. These key features are outlined in Figure 2.2, with population change divided into four stages.

The four key stage of the DTM are:

Stage I: Pre-transition, or pre-industrial societies, with high birth and high and fluctuating death rates; population growth is negligible and birth and death rates are roughly in balance.

Stage II: This stage is denoted by declining mortality, though fertility rates remain at pre-transition levels, thus resulting in rapid population growth.

Stage III: The fertility rate begins to decline, resulting in a slowing down of population growth. The timing between mortality decline in stage II and fertility decline in stage III is one of the key distinctive variables of the DTM.

Stage IV: At this stage populations are characterised by low fertility and low mortality, and fertility may fall below 'replacement levels', or may fluctuate slightly.

Probably the most interesting and useful question to ask about the DTM is why its popularity is not just restricted to the classroom. The model is more than a teaching device and has been a focal point for demographic debate for more than half a century. We will not review the content of these debates in detail here, as these are covered in depth elsewhere (see further reading at the end of this chapter). Rather this chapter aims to summarise the quintessential appeal of the DTM, and why it provides a rational and, at least at one level, thoroughly plausible explanation of demographic change that prioritises the importance of modernisation, declining infant mortality and individual prosperity as the main drivers of the demographic shift from large to small families.

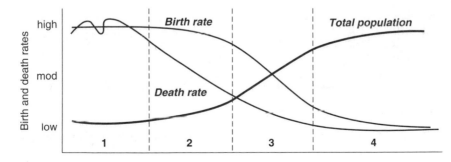

Figure 2.2   The Demographic Transition Model

To understand the key point of the demographic transition model and its endurance over time, we need to recognise the significance of how the model conceptualises the relationship between mortality and fertility decline, which assumes fertility decline as a response to earlier falls in mortality. The model, as first proposed by Notestein, was based on empirical data for Northern European countries. Historical time series of mortality and fertility data for countries such as Sweden, as illustrated in Figure 2.3, exemplify the hypothetical model, with a gradual overall decline in both fertility and mortality from the late eighteenth century onwards. However, declines in demographic events are not consistent, particularly for mortality as the decline in crude death rate is spiked by excessive death rates in some years.

While the model does appear to represent the historical experiences of countries such as Sweden, the usability of the model depends on two important features; first, the extent to which the model is replicated in other historical and geographical contexts, and second, whether the model is descriptive or explanatory, providing a causal link between declines in mortality and fertility. In other words, we are interested in whether the model just describes a possible relationship between fertility and mortality decline, or if there is an explanatory power to the observed relationship between mortality and fertility decline. These two conditions are clearly linked; if the model is explanatory, then we would expect a greater degree of replication in the relationship between fertility and mortality declines.

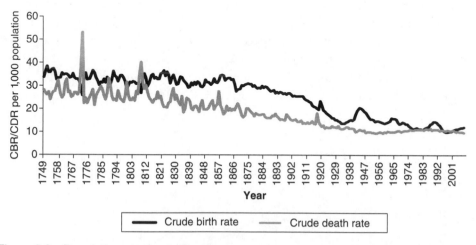

Figure 2.3   Population change in Sweden

*Source:* Statistics Sweden Statistical Database. Retrieved December 2010 from www.ssd.scb.se/databaser/makro/start.asp?lang=2

## Replicating the demographic transition model

David Reher's (2004) comprehensive analysis of the demographic transition as a global process compares the main characteristics of the transition for different times and places. Reher concludes that the historical experiences of European population change in the nineteenth and early twentieth century have implications for more recent population changes in Latin America and Asia as well as contemporary events in much of Sub-Saharan Africa. Using data collated by the UN, Reher identifies four main profiles of the demographic transition that are both time and place specific. Outlines of these four profiles are illustrated in Figure 2.4 (overleaf) and a description of each type is summarised in Box 2.2.

---

### Box 2.2   Reher's four profiles of the global demographic transition

- Forerunners: countries from Europe and North America.

Countries adhering to this profile depict the 'classic' demographic transition model though there are differences in rates of natural increase, with higher fertility in North America. In some Northern European countries (e.g. the UK, France and Sweden) declines in mortality and fertility had commenced by the late eighteenth century.

- Followers: countries in Asia and North America (especially countries with a European link either ethnically or politically).

In the forerunners there is a longer delay between declines in fertility and mortality resulting in more intensive population growth. A discernible increase in fertility prior to its decline is also evident (what some demographers have termed the 'ski jump' effect, see Dyson and Murphy, 1985).

- Trailers: countries in Asia, Africa and South America.

The profile for the trailers is broadly similar to that for followers, with the only consistent difference being the slightly later onset of fertility and mortality decline among the trailers.

- Latecomers: latecomers to the demographic transition are all found in Africa.

The classic demographic transition pattern is far less evident for the latecomers. There is a much greater gap between fertility and mortality decline, with fertility decline only discernible since the mid-1970s.

Source: Reher (2004)

---

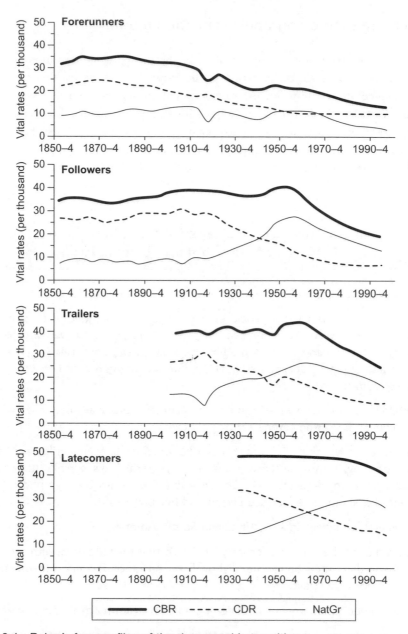

Figure 2.4    Reher's four profiles of the demographic transition

*Source:* Reher (2004). Redrawn with permission of the author

Reher's analysis illustrates how over time and place the applicability of the Demographic Transition Model varies quite considerably. In particular, in Sub-Saharan Africa the model appears to have less relevance, which raises questions about the universality of the Demographic Transition Model and its utility as an explanatory model.

Moreover, as Gould (2009: 90–1) describes, within Africa there are important national variations. The transition for some African countries has been shaped by conflict (for example in Liberia, Sierra Leone, Democratic Republic of Congo and Rwanda where genocide and civil war have halted declines in mortality and delayed any onset in fertility decline); and HIV/AIDS (for example in Zimbabwe where the AIDS epidemic has resulted in an increase in mortality since the early 1990s and, concomitant with declining fertility, has resulted in a reduction of population growth). Elsewhere in Africa, countries such as Ghana and Senegal have experienced a synchronisation in mortality and fertility declines, but there remain other countries where mortality decline has not brought about fertility decline. Gould describes this 'traditional' pattern as characteristic of some of the poorer countries across Africa from Mali and Niger in the West, to Angola and Chad in Central Africa and Uganda and Somalia in the East. Across Africa, therefore, the certainty of synchronisation of fertility and mortality decline has yet to be achieved. This suggests at least that the model as a descriptive *and* an explanatory tool might not be universal.

## Demographic transition model as an explanatory model

The key feature of the demographic transition model is the delayed synchronisation between mortality and fertility decline, thus if the model is to have any explanatory power, then we need to consider how a decline in mortality may be causal for a decline in fertility. Early adherents to the Demographic Transition Model, such as Kingsley Davis, regarded mortality declines as 'both a necessary and a sufficient stimulus for fertility decline' (Cleland, 2001: 60). Yet in more recent years this causality has been called into question. In classic interpretations of the Demographic Transition Model, the causal link between mortality and fertility was associated with three main mechanisms: physiological, replacement and insurance (Cleland, 2001).

The physiological mechanism is the most straightforward and links infant death with shortened birth intervals associated with curtailed breast-feeding. This is because when a baby dies the mother will stop breast-feeding, and as breast-feeding suppresses ovulation (this known as lactational amenorrhea and is discussed in more detail in Chapter 5) she is more likely to conceive another baby within a short period of time, than if the baby had survived and the mother breastfed for a longer period. If this mechanism is effective, high infant mortality will be associated with shorter birth intervals. As infant mortality declines and more babies survive the first years of life, birth intervals will extend thus reducing overall fertility.

The other two mechanisms linking infant mortality and fertility depend on couples adjusting their fertility in response to an infant death. For the replacement hypothesis, it is assumed that couples will replace an infant death quickly with another child, thus as infant mortality declines this replacement effect will be dampened. The insurance mechanism has proved to be the most intangible in terms of empirical evidence (Cleland, 2001), and is based on the assumption that as infant mortality declines, couples can reduce their fertility as they can guarantee survival of an heir from a smaller family.

The problem in trying to establish any of these direct causal links between mortality and fertility decline is, as Cleland concludes, that there are just too many mediating factors through which mortality decline can be causal for fertility decline. For this reason, many demographers have assumed that the explanatory power of the DTM is limited. Yet as Cleland notes, failure to identify a direct causal link between mortality and fertility decline does not necessarily infer that the DTM is merely descriptive, rather, we can recognise that mortality decline remains the 'common underlying cause of developing-world fertility transition – chronologically remote but nevertheless fundamental' (2001: 87). However, the DTM on its own does not provide a coherent and unique explanation of fertility and mortality decline.

## Mortality decline

The DTM demonstrates that it is possible for societies to escape the Malthusian trap and achieve population growth as a result of continued decline in mortality and increases in productivity. Yet there is considerable disagreement among scholars as to how this was achieved in the past and, as a consequence, how it might be realised in societies that remain in the early stages of the demographic transition. Context is clearly important; we might expect very different causal explanations of mortality decline in nineteenth-century Sweden, compared to contemporary Sub-Saharan Africa. In particular, the relative contribution of medical innovation, income and environmental factors will be context specific. The opportunities for medical innovation to bring about mortality decline were clearly greater in the latter half of the twentieth century compared to the nineteenth century. However, there are a number of competing theories that seek to provide a universal explanation of mortality decline that prioritise specific trigger mechanisms that can bring about a fall in death rates. One way of conceptualising these different theories is whether mortality decline is exogenous or independent of social and economic change. In posing this question, population researchers are considering the extent to which mortality decline can be achieved through improvements in healthcare and knowledge of how to reduce the transmission of infectious disease, or if mortality decline is dependent on improvements in economic and social wellbeing.

Thomas McKeown's (1976) analysis of mortality decline is one of the most cited but also one of the most debated explanations of mortality decline. McKeown attributed the 'modern' rise in world population from the 1700s onwards to social and economic improvements rather than more targeted medical or public health interventions. In particular, McKeown identified the decline of tuberculosis mortality in nineteenth-century Europe as occurring prior to any medical advances (the bacillus for tuberculosis was not identified by Koch until 1882, and mass vaccination did not start until after the Second World War). The enduring interest in McKeown's thesis reflects the significance of his key question: are targeted public health interventions more effective than more broad-based improvements in living standards in bringing about a decline in mortality?

Though McKeown's empirical analysis focused on historical declines in mortality, this question is clearly relevant for contemporary developing societies and has important implications for development policies: should aid be targeted through public health initiatives or through more general social/economic development programmes? McKeown's judgement that broad based secular improvements in living standards are more effective in bringing about mortality decline than targeted intervention has remained a point of contention for scholars of both historical demography and more recent population change. At issue here is a key question that the demographer Samuel Preston (2007) has asked: is it possible to achieve low mortality without sustainable economic development? McKeown's response would be an unequivocal no, that mortality decline is brought about by substantial improvements in nutritional status caused by rising living standards. McKeown's conjecture is assumed by other theorists, most notably Malthus, for whom, as discussed earlier, a negative relationship between income and mortality was a key tenet of his treatise on population.

McKeown's thesis is not universally supported. For example, Jack Caldwell (1986) provides a very different answer to Preston's question, that low mortality can be achieved in poor societies, though it requires effective health care interventions, particularly those that target infant and child health. Caldwell suggests that 'the constraint of material resources can be very largely overcome, at least in the contemporary Third World, which can import both medical technology and social institutions' (1986: 176). Drawing on the experiences of three countries, Sri Lanka, India (Kerala), and Costa Rica, that have achieved unusually low mortality through a combination of political and social will, Caldwell identifies the following conditions that need to be met for mortality rates to be lowered:

- sufficient female autonomy;

- commitment to both health services and education, with female schooling on a par with male schooling;

- accessible health services for all regardless of location or income;

- efficient health services;

- provision of a 'nutritional floor' or distribution of food;

- universal immunization;

- effective antenatal and postnatal health services (Caldwell, 1986: 208).

It would seem reasonable to conclude that the relative importance of rising living standards will not be the same for different populations. In particular, the relationship between rising income and life expectancy will vary over time, with McKeown's argument more applicable to mortality decline in developed countries during the nineteenth century, while factors exogenous to a country's wealth have proven to be more relevant in more recent years.

    This relationship is empirically demonstrated by Samuel Preston in a seminal paper first published in 1975. Preston's analysis of the relationship between income and life expectancy at three time periods (1900s, 1930s and 1960s) is illustrated in Figure 2.5. The two trend lines plotted for the 1930s and 1960s demonstrate how the relationship between life expectancy and income has shifted upwards during the middle part of the twentieth century – that is, the same level of GDP is associated with a higher life expectancy in 1960 than in 1930. Thus Preston concludes, 'factors exogenous to a nation's level of economic development have affected the level of mortality' in both the 1930s and 1960s. In particular, he argues that while this might be more apparent if we consider the relationship between life expectancy and income in the 1960s, commentators such as McKeown have underestimated the impact of health technology and more direct intervention for earlier periods. For example, the decline of diseases such as plague in the seventeenth and eighteenth centuries owed as much to changes in housing as to increases in income and living standards. Mortality decline, it would appear, cannot be readily explained by mono-causal theories.

## Fertility decline

As with mortality, a key debate in explanations of fertility decline concerns the importance of economic factors. That is, is the best way to achieve fertility decline through economic prosperity? There is undeniably a clear negative relationship between fertility and income and this holds for different scales: if we are looking at individuals, populations or sub-populations, those with higher income have lower fertility. For example, Bongaarts' (2006) analysis of demographic and health survey data for 38 countries in the 1990s illustrates a reasonably strong fit between level of fertility (as measured by the total fertility rate) and GDP per capita; that is, countries with higher GDP had lower fertility. Moreover, in societies in Stage III of the demographic transition, wealthier individuals are more likely to be innovators of fertility decline. One possible explanation for this relationship is that the opportunity costs of having children increase with prosperity

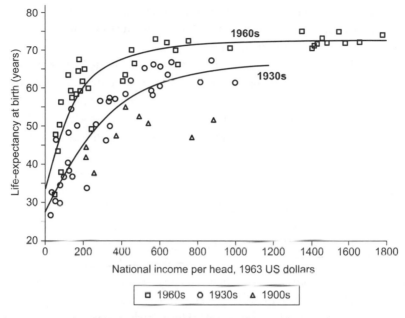

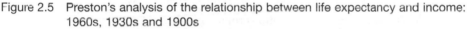

Figure 2.5    Preston's analysis of the relationship between life expectancy and income:
1960s, 1930s and 1900s

*Source:* Preston (2007). Redrawn with author's permission

(i.e. the costs of providing for children increase as children are no longer, or less, economically active and engaged in schooling for longer), and wealthier couples are more
likely to recognise these costs and are thus motivated to reduce their fertility.

One of the most widely cited interpretations of the relationship between income and
fertility is Caldwell's theory of wealth flows (1982). Caldwell argues that in pre-transition,
high fertility societies the generational wealth flow is from children to parents; that is,
it is children who contribute to family prosperity. However, in transitional low fertility societies parents are expected to provide for their children's economic well-being.
Caldwell and colleagues have used this framework to explain the persistence of high
fertility in West Africa (Caldwell and Caldwell, 1987). They argue that the pronatalist
culture of African societies, particularly in rural communities, is associated with the
persistence of upward generational wealth flows. Put simply, large families make economic sense in rural Africa, where family workers can contribute positively to household income. In order for fertility to decline, Caldwell argues that this flow needs to
be reversed. In more recent years, Caldwell (Caldwell et al., 1992) recognises that
some fertility decline has been achieved in Africa, and suggests that this is a result of
urbanisation and increased access to contraception.

Education is also a key factor for fertility decline. As Gould points out, the gap
between fertility levels in urban and rural Africa is associated with education differentials. More highly educated women have a better understanding of contraception, can

afford it and are more likely to use it effectively. For example, in Kenya in 2003, 51 per cent of women with at least secondary education used contraception compared to 8 per cent of women with no education (Gould, 2009: 145).

At this point, it is germane to reflect on the overall tone of demographic research on fertility that very much assumes the desirability of fertility control, and this assumption underpins many national and international population programmes. One way of consider-ing this issue is to reverse the question: declines in fertility are associated with the exten-sion of education, improvements in women's and children's health and female autonomy (to name just three factors), so is sustained high fertility, therefore, a barrier to these social advances? Yet if fertility decline is to be achieved without direct intervention, for example the introduction of draconian anti-natalist policies, the benefits of smaller family size need to be both recognised and achievable for all sectors of society. The American demographer Ansley Coale (1973 quoted in Cleland and Wilson, 1987: 6) identified three conditions for fertility decline: that fertility is 'within the calculus of conscious choice'; effective mechanisms of reducing fertility must be known about and available; and couples must recognise the perceived advantages of lower fertility. However, as Cleland and Wilson (1987) conclude, far more attention has been given to the latter condition rather than the first two, and as such economic theories of fertility decline have generally shaped both academic debate and political interventions round fertility.

Thus a common assumption of fertility decline is that fertility decisions, whether for fewer or more children, are essentially rational. The limitations of economic rationality have, however, been explored by several commentators. David Kertzer's (1997) detailed study of historical Italian populations challenges the assumption that high fertility was desired by couples in pre-transitional historical populations. Research on European fertility decline carried out at Princeton University, led by Ansley Coale (Coale and Cotts Watkins, 1986) identified the importance of cultural and linguistic diffusion in the spread of fertility decline. Their detailed statistical analysis of regional European fertility declines revealed the difficulty of indentifying key relationships particularly between indicators of modernity such as occupation and education and fertility. The Princeton project, rather than providing a unifying theory of fertility decline, hints more towards ambiguities and vague explanations. This suggests the importance of knowledge and what Cleland and Wilson (1987) term an 'ideational' shift in populations that brings about fertility decline. Cleland and Wilson argue that the empirical evidence of links between economic structures and fertility are actually quite weak, while those between culture and education with fertility are much stronger (Bongaarts, 2003). They argue that cultural values and education promote the initial acceptance of new ideas that can lead to a quick uptake of birth control. They also suggest that Caldwell's theory of wealth flows is essentially an ideational theory; what matters in this theory is a shift in family morality that encourages parents to be aspirational for their children. It is this shift that turns children from economic providers to consumers. It is not possible there-fore to isolate one overriding causal factor that will bring about changes in fertility, as structural factors, and particularly economic variables, will be mediated by cultural practices (Caldwell, 1997).

The main conclusion from this review of scholarship on mortality and fertility decline would caution against any attempt to derive an overarching theory of demographic change. Both mortality and fertility decline can be brought about by very different combinations of factors. Yet despite the research on population change that has done much to challenge the certainties of the DTM, the model itself has not been discredited. An explanation for its persistence relies as much on optimism, that the idea of the inevitability of mortality and fertility decline brought about by economic modernisation is intrinsically appealing (see for example Cleland, 2001). The DTM also raises important questions about how development is achieved and the inter-relationship between demographic trends and social, economic and political development. In a recent review Tim Dyson (2010) argues that the transition should be regarded as a self-contained process and though clearly inter-related with other social, economic and political processes, it is not dependent on these. Moreover the decline of mortality facilitates other social changes, in particular in societies with lower mortality life is more secure and stable, and thus has implications for politics and social relations. As we explore in Chapters 7 and 8, individuals may become less dependent on wider kinship networks and, as such, horizontal kinship ties are weakened while vertical ones become more significant. Dyson's claim about the independence of the transition raises important questions about the relationship between population and development (see Canning, 2011 for a review of Dyson's argument) but also highlights how the study of population change involves more than a description of demographic processes, as changes in mortality, fertility and population size have profound implications for economic, social and political processes.

# Other transitions

The DTM is the best-known example of a transitions approach to demographic research, though the same idea, that demographic events move through discrete stages associated with societal change, has been applied to other population parameters, specifically epidemiology and migration.

## Epidemiological transition

The Epidemiological Transition Model (sometimes known as the Health Transition Model) is immediately associated with the DTM as it examines the contributions of different causes of death associated with modernisation and provides one explanation for the broad trends in life expectancy described above. This theory proposes that for the poorest countries at an early stage of economic development the main causes of death are parasitic, infectious and nutritional which account, in particular, for high rates of infant mortality. Through the process of economic development deaths associated with these diseases fall due to improved diets, vaccination programmes and health services. Non-communicable diseases such as circulatory diseases and cancers, which tend to affect the adult and oldest ages, become predominant.

Table 2.1 illustrates the main causes of death comparing low and high income countries in 2004. It is clear that low income countries have more deaths from HIV/AIDS, infectious and diarrhoeal diseases. All the other regions of the world have moved away from these types of deaths so that non-communicable diseases comprise the majority of the deaths, namely heart disease, cerebrovascular diseases and cancer.

The epidemiological transition encapsulates the transition from high mortality to low mortality conditions and how this relates to the main causes of death. The main characteristics of low and high mortality societies are outlined in Box 2.3.

As with the DTM, the epidemiological transition can be either explanatory or descriptive. The requirement for the epidemiological transition to be explanatory is that there is a causal relationship between cause of death and income levels. The richest countries went through the epidemiological transition in the early part of the twentieth century whilst for the middle income countries the transition occurred from the 1950s (Gould, 2009). However, it is important to note that the mortality statistics in Sub-Saharan Africa do not imply that these countries will simply follow the richest and middle income countries through the epidemiological transition. The five conditions that are responsible for most of the child mortality in poor countries are, in the most part, easily treatable. This makes the Millennium Development Goal to tackle child mortality in these countries particularly salient.

Table 2.1   Global mortality: main causes of death 2004 (total deaths in millions)

| Disease or injury | Total deaths | Disease or injury | Total deaths |
|---|---|---|---|
| **Low income countries** | | **High income countries** | |
| Lower respiratory infection | 2.9 | Ischaemic heart disease | 1.3 |
| Ischaemic heart disease | 2.5 | Cerebrovascular disease | 0.8 |
| Diarrhoeal diseases | 1.8 | Trachea, bronchus, lung cancers | 0.5 |
| HIV/AIDS | 1.5 | Lower respiratory infection | 0.3 |
| Cerebrovascular disease | 1.5 | COPD | 0.3 |
| COPD | 0.9 | Alzheimer's and other dementias | 0.3 |
| Tuberculosis | 0.9 | Colon and rectum cancers | 0.3 |
| Neonatal infections | 0.9 | Diabetes mellitus | 0.2 |
| Malaria | 0.9 | Breast cancer | 0.2 |
| Prematurity and low birth weight | 0.8 | Stomach cancer | 0.1 |

Source: WHO (2008) The Global Burden of Disease: 2004 Update Geneva: WHO, Table 2, p. 12

---

**Box 2.3   Characteristics of high
and low mortality societies**

High mortality societies

- Low income

- Communicable disease

  o Waterborne infections

  o Airborne infections

  o Contagious infections

- High periodicity

- Endemic diseases (e.g. malaria)

- 'Age of pestilence and famine'

- Disproportionally affects young

Low mortality societies

- High income

- Heart disease, stroke and cancer

- Disproportionally affect older ages

---

For the richest countries, the focus of health professionals is now turning to diseases associated with poor lifestyle choices. Bans on smoking in public places have been introduced in several European countries and there are increasing pressures to clearly label food in terms of its nutritional contents. Such policies are intended to 'nudge' people away from the kinds of behaviour that lead to conditions such as circulatory disease and lung cancer in later life. The extent to which governments should attempt to pursue such policies that attempt to influence the behaviour of their citizens is debated, with some voices arguing that such an approach violates personal freedoms.

The Epidemiological Transition Model does not provide a complete explanation for the health transition in all cases and it is worthwhile considering other factors that influence societal levels of mortality. As discussed above, there are countries that have experienced mortality decline which was not preceded by particularly high levels of economic development. In such situations it is suggested that colonial governments or international organisations, such as the UN, may have contributed to the declines in mortality through the imposition of medical and public health programmes. Climatic factors also influence levels of mortality – countries with warmer climates are more prone to diseases such as diarrhoea or malaria. Moreover, climate change has been cited as a reason for increased prevalence of malaria in parts of Sub-Saharan Africa in recent years, particularly highland areas which were previously malaria free (Hay et al., 2002). Additionally, the political stability of a country influences the health of its population. The policies of land re-allocation in Zimbabwe are thought to have contributed towards shortages of food and the break-up of the Soviet Union has led to falling life expectancies amongst males that have been linked to alcoholism (Shkolnikov et al., 2001). However, in all of these contexts it is not possible to establish a unique explanation for the epidemiological transition, or where the transition has been reversed. Rather, as discussed in relation to overall decline in mortality for the DTM, the causal mechanisms of mortality change come about through the complex interrelation of medical, economic, social and political factors.

## Migration transition

One important omission from the DTM is migration, the third component of population change. However, an equivalent to the DTM and ET was proposed for understanding the association between countries' economic development and their migration experience by Zelinsky. Zelinsky's (1971) Hypothesis of Mobility Transition combines 'laws of migration' (Ravenstein, 1885, 1889; see Chapter 6) with the Demographic Transition Model. He hypothesised that 'there are definite, patterned regularities in the growth of personal mobility through space-time during recent history, and these regularities comprise an essential component of the modernization process' (Zelinsky, 1971: 221–2). The basic proposition of the model is that as economic development and wealth increase, so, too, does migration. Zelinsky proposed that the Mobility Transition Model can be applied to different forms of mobility including international, internal and circulatory (these are discussed more in Chapter 6). The model is based on a series of hypotheses, three of which are:

- A transition from a relatively sessile condition of severely limited physical and social mobility toward much higher rates of such movement always occurs as a community experiences the process of modernization.

- For any specific community, the course of the mobility transition closely parallels that of the demographic transition and that of other transitional sequences.

- Such evidence as we have indicates an irreversible progression of stages.

The Mobility Transition Model proposed by Zelinsky has five phases which are summarised in Table 2.2 (overleaf). These can be read in a comparable way to the phases of the Demographic Transition Model. In particular, they present a model of changes in the levels and characteristics of migration experienced by a country or region as it develops from a pre-modern traditional society to a super-advanced society. Zelinsky suggests a peak of migration for early transitional societies, levelling off for advanced societies and declining for super-advanced societies due to increases in communication and controls on migration.

Zelinksy's idea has been widely applied and has been shown to describe differences in population redistribution between more and less developed countries. However, as with other general models, including the DTM, questions can be asked about how applicable the model is to different times and places; and whether it has descriptive and/or explanatory power (Skeldon, 1997). For example, de Hass (2010) suggests that the model should be refined to take into account potential stagnation and reversal in the transitions; and that opportunities and aspirations will affect migration as well as income differentials. The hypothesised decline in migration for super advanced societies can also be questioned – it can be argued that mobility has become more rather than less pertinent in advanced societies (for example, see Urry, 2007) – as can the prediction of urban focused moves which fails to recognise trends of counter urbanisation (see Chapter 6; Champion, 1989). More generally, critiques point towards understanding migration as a process experienced by individuals as well as societies and places, for example by taking a life course perspective (see below) or a humanist perspective (see Chapter 6).

## Why are transitions important? From aggregate to individual behaviours

The study of demography and of the transitions in particular has tended to use the nation state as *the* unit of analysis. The focus of the DTM and other related transitions is towards aggregate population change and understanding how fertility and mortality change is experienced collectively rather than individually. However, this is not to say that we should ignore the issues of scale and boundary formation.

If we take the issue of boundaries first, the essence of the DTM is that population change is not uniform, but that it is experienced at different rates and at distinct stages depending on where you live. The early pioneers of fertility decline in Northern Europe experienced a very different DTM compared to later comers in Asia or Africa. However, we should also be cautious in reifying the nation state in demarcating the limits to fertility and mortality decline. Clearly political boundaries will contribute to demographic changes, particularly regarding the introduction of policies to promote

Table 2.2　Characteristics of the five phases of Zelinsky's hypothesis of mobility transition

| Phase of transition | Level of migration | Rural–urban direction | Colonisation | Emigration and immigration | Circulation | Other characteristics |
|---|---|---|---|---|---|---|
| 1: The Pre-modern Traditional Society | Low | Not relevant | Not relevant | Not significant | Low, linked to land utilization, social visits, commerce, warfare, or religious observances | |
| 2: The Early Transitional Society | Very high | Massive rural to urban movement; creation of new cities | Movement to colonisation frontiers where profitable land is available | Major outflows and some immigration of skilled migrants | Significant growth in circulation | |
| 3: The Late Transitional Society | High | Continued rural to urban movement | Lessening flow of migrants to colonisation frontiers | Declining or ceased emigration | Further increases in circulation, with growth in structural complexity | |
| 4: The Advanced Society | High: level and oscillating | Continued but reduced rural to urban movement accompanied by vigorous movement from city to city and within individual urban agglomerations | Settlement frontiers stagnant or retreating | Significant net immigration of unskilled and semiskilled workers from relatively underdeveloped areas. Potential significant international migration of skilled and professional people | Vigorous accelerating circulation, particularly the economic and pleasure-oriented | |
| 5: A Future Superadvanced Society | Declining (due to better communication and delivery systems) | Almost entirely inter- and intra-urban | | Some further immigration of relatively unskilled labour | Acceleration in circulation | Strict political control of internal as well as international movements may be imposed |

Source: Adapted from Zelinsky (1971)

contraception or health interventions that may vary from nation state to nation state. Yet, as discussed above, the classic DTM assumes a closed demographic system that makes no reference to migration, and this is an important omission, as in practice the forces that drive the DTM model are not necessarily contained by national boundaries. As historians of fertility decline in Europe have documented, fertility control was discernible at subnational, regional and even city scales and diffusion of new practices was stimulated by cultural and economic networks (see Lesthaeghe and Neels, 2002 and Brown and Guinnane, 2002). Regional diversity is important in contemporary contexts – for example, in Africa urbanisation is recognised as a key condition for sustained fertility decline. Regional dynamics and diversification are therefore important conditions for demographic change.

Yet even if we recognise the importance of both regional and transnational change, we should also be cautious in assuming that the DTM will be experienced uniformly within a city, region, nation, or global region. Put simply, the DTM does after all depend on changes in individual behaviours. This is most apparent when considering fertility decline, but is also relevant for aspects of mortality decline, particularly regarding measures that individuals themselves can take to increase life expectancy, such as methods of infant feeding as well as other aspects of personal hygiene. Thus as David Reher observes, drawing on the work of the Belgium demographer Ron Lesthaeghe: 'Everywhere the demographic transition basically removed reproductive behaviour from the realm of social control and placed it within the purview of individual choice' (Reher, 2004: 34).

We shall return to consider the significance of choice in Chapter 8 and how it shapes contemporary fertility and family formation practices. For now our purpose is to appreciate that the DTM is not just a descriptive model of aggregate changes in fertility and mortality, but that these changes signify more fundamental shifts in how individuals seek to control their fertility, and recognise that this control is in their remit. The DTM therefore incorporates important and irreversible social changes that have implications beyond demographic processes. The DTM can be read as a representation of the changing relationship between self and society. While the model is concerned with national, aggregate populations, it is individual behaviour that brings about the demographic change.

## Life course approaches

In recognition of the agency of individuals to shape their own lives, including in the realms of fertility, health and migration, studies of social change in demography and geography (and other disciplines), have, over the last few decades, developed theories and methods to understand decisions and experiences through a person's life (Hunt, 2005). A life course approach is concerned with the sequence of events (and processes) that a person experiences, and key transitions for that individual, such as leaving home,

beginning employment, partnership, having children, retirement and death of a partner. It is interested in how these key transitions through the life course are shaped and
experienced; how they affect each other; and how they are associated with aspects of
identity including gender, class and ethnicity. Thus, life course studies of demographic
change are rather different from aggregate level understandings of societal transitions
presented by the DTM, the epidemiological transition and the migration transition
models. Life course approaches enable both description and explanation, for example
about the experience and effect of age, generation and period on life experience.

Examples of life course understandings of population change can be found in this
book. For example, Chapter 6 considers individual as well as structural (societal) factors
that influence migration and introduces a life course perspective on migration; and Chapter 8 discusses how individuals make choices about family formation (partnership and
children) and introduces the concept of individualisation which places emphasis on individuals as the agents of change in their own lives.

## Summary

This chapter has reviewed the concept of transition and its centrality in population studies, with particular reference to the Demographic Transition Model (DTM). The chapter can be summarised by the following points:

- The DTM provides a framework to conceptualise the relationship between mortality
  decline and fertility decline, and how this results in population growth, before an
  equilibrium between low fertility and mortality is achieved.

- The DTM was originally developed using empirical case studies of demographic
  change, particularly those of Northern European countries in the eighteenth and
  nineteenth centuries. While the general principles of mortality decline followed by
  fertility decline can be applied to other contexts, especially other global regions at
  different times, important regional variations in the model are discernable. These in
  particular relate to the timing of the lag between mortality and fertility, and even if
  the former is a necessary condition of fertility decline.

- Debates about whether the DTM is an explanatory or descriptive model hinge on
  understanding the relationship between mortality and fertility decline, and the
  extent to which the former brings about the latter. A key concern here is to unravel
  the relationship between economic development and mortality and fertility decline.
  Here the evidence is inconclusive, in that while a general causality between income
  and lower rates of fertility and mortality can be identified, there are case studies of
  countries or regions experiencing population change without necessarily benefiting
  from economic modernisation.

- The concept of transition is applicable to other demographic processes. The epidemiological transition summaries the relationship between level of mortality, cause of death and income. The general thesis is that countries with high levels of mortality experience higher rates of communicable disease with higher periodicity, while in low mortality societies the majority of deaths are from non-communicable diseases such as heart disease and cancer.

- The Hypothesis of Mobility Transition suggests that, in parallel to the Demographic Transition, economic development of a country (or region) is accompanied by higher levels and diversity of mobility, which stabilise as the country reaches advanced stages of economic development. The mobility transition, and its underlying assumption that differential migration is associated with differential income or wealth, has been widely applied but questions have also been raised about its applicability spatially and temporally, and the impact of non-economic drivers of migration.

- While the concept of transition in population studies focuses on aggregate population change, the working of the models depends on individual action, that is population change is brought about, in part, by individual behaviour. Life course approaches focus on behaviour and experience through transitions in individuals' lives. Unravelling the relationship between individual and aggregate behaviour is one of the key challenges in population research.

# Recommended readings

## Pre-transition populations

For a review of population history over time, see Livi-Bacci, M. (2001) *A Concise History of World Population*, 3rd edn. Oxford: Blackwell.

## Demographic Transition Theory

The DTM is discussed in depth in a number of essential readings on demography and population studies. Key contributions include John Cleland's review of the relationship between mortality and fertility decline: Cleland, J. (2001) 'The effects of improved survival on fertility: a reassessment', *Population Development Review*, 27 (1): 60–92. Van de Kaa provides a detailed review of scholarship on fertility decline: Van de Kaa, D.J. (1996) 'Anchored narratives: the story and findings of half a century of research into the determinants of fertility', *Population Studies*, 50 (3): 389–432. Dudley Kirk and David Reher both provide an assessment of how the DTM has been achieved in different regional and temporal contexts: Reher, D. (2004) 'The demographic transition revisited as a Global Process', *Population Space and Place*, 10: 19–41; and Kirk, D. (1996) 'Demographic

Transition Theory', *Population Studies*, 50 (3): 361–87. Tim Dyson's recent review that situates the demographic transition with reference to modernity and development is *Population and Development* published in London by Zed Books in 2010.

## Mortality decline

Thomas McKeown's classic account of the decline of mortality is *The Modern Rise of Population* published in 1976 by Arnold in London. Jack Caldwell's research is summarised in 'The routes to low mortality in poor counties' published in 1986 in *Population and Development Review*, 24 (4): 171–220. Samuel Preston's critical discussion of the relationship between mortality decline and income is outlined in 'The changing relation between mortality and level of economic development', which was reprinted in 2007 in *International Journal of Epidemiology*, 36: 484–90.

## Fertility decline

Caldwell's theory of fertility decline is presented in his 1982 book *Theory of Fertility Decline* published by Academic Press of New York. There are numerous publications from the Princeton project, though an overview is provided in: Coale, Ansley J., and Watkins, Susan Cotts (eds) (1986) *The Decline of Fertility in Europe*. Princeton, NJ: Princeton University Press. Cleland and Wilson's view is presented in their 1987 essay: 'Demand theories of the fertility transition: an iconoclastic view', *Population Studies*, 41 (1): 5–30. For a review of fertility decline in developed contexts see John Bongaarts's 2002 essay 'The end of fertility transition in the developed world', *Population And Development Review*, 28 (3): 419–43.

## Other transitions

For a review of the epidemiological transition see: Gatrell, A.C. and Eliott, S.J. (2009) *Geographies of Health: An Introduction*. Oxford: Blackwell. Zelinsky's mobility transition was first outlined in his paper: Zelinsky, W. (1971) 'The hypothesis of the mobility transition', *Geographical Review*, 61 (2): 219–49.

# 3

# POPULATION PROJECTIONS AND DATA

The Demographic Transition Model (DTM) presents a paradigmatic perspective of population change that has shaped population research for over half a century. In particular, the quest for universal explanations of population change that can be replicated in different contexts has been central to both scholarship and international development programmes and many demographers have contributed directly to World Bank, UN and other international agencies' agendas and programmes. Moreover, the DTM continues to shape international interpretations of population change. But population research has not stood still for over half a century; while the DTM paradigm remains important, population researchers have access to more varied data at different geographical and temporal scales that make the idea of a single unified approach harder to sustain. The DTM emerged at a time of limited demographic data availability which necessitated analysis at larger geographical and temporal scales, i.e. analysis of countries over long time periods. Yet as data become both more reliable and available, population researchers are able to analyse population dynamics at finer geographical levels, with emphasis on diversity both within and between geographical regions. In this chapter we review how population research has progressed from the transition paradigm to focus more directly on diversity and destandardisation with reference to the dynamics of population projections. The chapter then outlines the main sources of demographic data and considers issues of data access and reliability.

The chapter addresses the following questions:

- What are the assumptions of global population projections and to what extent do these rely on the DTM?

- What do population projections tell us about the dynamics of future population change?

- Where are population data obtained from?

- What geographies are used in population research?

## Population projections

In the introduction to this book we considered how early attempts to project population have in time proven inaccurate. Nowadays it is reasonable to conclude, following Lee, that:

> demographers have been quite successful in their population forecasts, well represented by the biennial UN population projections for countries, regions, and the world population. (Lee, 2011: 569)

Moreover, the publication of the UN projections is now well covered in the media, particularly when important population milestones are reached, as the extensive media coverage of 7 billion people in 2011 demonstrated. Yet while milestones are important in raising awareness of population size, what is really at issue here is how to understand the underlying rate of population growth (or decline) and how this might change in the future. In a major review of population projections published by the National Research Council in 2000 (Bongaarts and Bulatao, 2000) the editors highlighted in their introduction that the underlying dynamics of population projections are often misunderstood. For example, the authors compared the 1996 UN projections with those from 1994; a small adjustment to projected annual growth resulted in the projected population for 2050 being smaller in 1996 than 1994 by 466 million people. Popular interpretations at the time were that future population growth was not as big a problem as had been originally proposed, yet the difference between the two sets of projections is within the range of future global population, in other words the 466 million can be accounted for by probabilistic adjustments in each projection. Too often we take population projections at face value without considering the assumptions that are essential for population projections.

UN population projections for the current century have received considerable coverage, and you are probably familiar with projections that put the future global population at 9, 10 or 15 billion in the year 2050. Not surprisingly there is no unanimously agreed projection of future population; rather we need to consider a range of possible outcomes. The UN projections for 2010 give four possible future trends based on different assumptions for future fertility (the future course of mortality is assumed to be the same in all variants, as is migration), and these are illustrated in Figure 3.1.

The constant fertility variant which suggests a future global population of over 25 billion by the end of the century assumes that for each country fertility remains

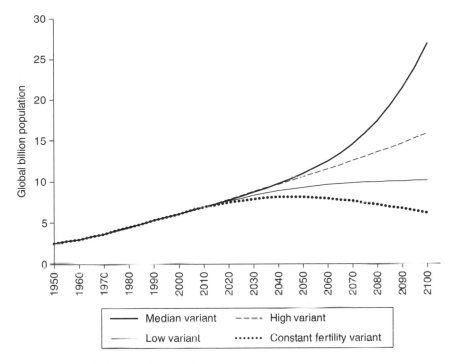

Figure 3.1    UN global population projections, 1950–2100

*Source:* UN Department of Economic and Social Affairs (no date) World Population Projections 2010 Revision http://esa.un.org/unpd/wpp/unpp/panel_population.htm

constant at the level estimated for 2005–2010. However, it is reasonable to assume that there will be some fertility decline, particularly if all countries are eventually to pass through all stages of the DTM. The medium variant, which might be interpreted as the 'best guess' of future population, is based on empirical fertility trends estimated for all countries of the world for the period 1950 to 2010 (UN no date). The UN method assumes that fertility in all countries will pass through three broad phases based on the DTM:

i    a high-fertility pre-transition phase;

ii    the fertility transition itself; and

iii    a low fertility post-transition phase during which fertility will probably fluctuate around and remain close to replacement level (or converge towards it) (page 2).

The medium variant projection models how fertility in each country passes through these stages (or remains in Stage IV for countries that have completed the DTM). The

UN projections also include two variants on this medium projection: the higher variant assumes that fertility is greater by 0.5 births per women compared to the medium variant, while the lower variant allows for fertility to be lower by the same margin. The difference between the high and low variant projections is 2.5 billion people by 2050; yet both projections assume some process of fertility transition. Clearly these differences are more than just trivial, but highlight the uncertainly of projections, as the National Research Council summarized in 2000, based on earlier UN projections:

> We estimate that a 95-per cent prediction interval for world population in 2030 would extend from 7.5 to 8.9 billion, and a similar interval for world population in 2050 would extend from 7.9 to 10.9 billion. The intervals are asymmetric around the U.N. medium projection of 8.9 billion in 2050. This indicates that, based on the record of previous projections, a greater risk exists of a large understatement of future world growth than of a large overstatement. (Bongaarts and Bulatao, 2000: 10)

However, the group of demographers who wrote the 2000 report were unanimous about one prediction: that a *decline* in world population between 2000 and 2050 is extremely unlikely, even though the magnitude of future growth cannot be known exactly. Moreover, the UN projections are based on the assumption that the fertility transition is inevitable and can be achieved in all countries, even in those countries, which are now almost exclusively in Sub-Saharan Africa, that have not yet entered the fertility transition stage of the DTM.

If projecting the future of global population growth is problematic, the magnitude of variation in projections is often greater at small geographical scales. This is because errors in projecting population in nations can cancel each other out when they are amalgamated. Again the National Research Council review discusses the importance of scale:

> Across 13 large countries, the median prediction interval for population in 50 years runs from 30 per cent below the point forecast to 43 per cent above it, for a total width of 73 per cent. This width is more than three times the width of the corresponding projection interval for world population. (Bongaarts and Bulatao, 2000: 10)

The range of potential population at a national level is illustrated in Figure 3.2 which plots the UN variants for three countries: China (which currently has the largest population hence is greatly significant for global population, though UN median variant projections predict that India's population will surpass that of China in 2025); Nigeria (the country with the largest population that is at the very early stages of the fertility transition;) and the UK.

These graphs give the three UN variants (medium, high and low) and the 85 per cent and 95 per cent confidence intervals for the medium variant. These figures illustrate the magnitude of the differentials not only within but also between countries. As can be seen in the graphs, while the low variant projection leads to an eventual decline of population

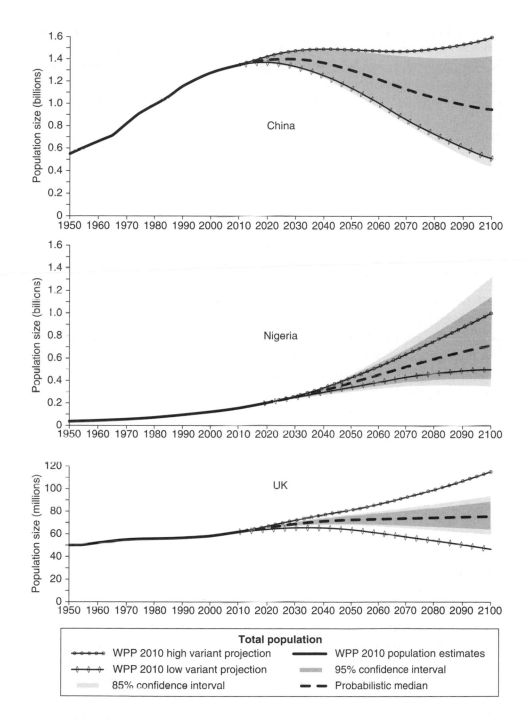

Figure 3.2    Population projections to 2100: China, Nigeria and the UK

*Source:* UN World Population Projections 2010 Revision http://esa.un.org/unpd/wpp/unpp/panel_population.htm

by the end of the century in China and the UK (in both countries back to the population size observed in the 1950s), in Nigeria all three projections predict population
increase. The differential between the higher and lower variants in China is more
marked than in Nigeria (for China there is a difference of 1 billion people between the
two extreme variants in 2100). However, the grey shadings, which indicate the confidence interval for the medium variant, are much greater in Nigeria compared to the
UK. This is not surprising as the UK has completed the fertility transition so it is much
easier to model future fertility with some certainty, compared to Nigeria where the
magnitude of future fertility decline is not known. It is interesting to observe that if
Chinese fertility falls to levels between the median and lower variants and Nigeria's
follows the higher variant, Nigeria's population is projected to be greater than that of
China by the end of the century. Thus if we turn our attention from global population
growth to individual countries, not only is the degree of uncertainty enhanced, but the
possibility of population decline in some parts of the world is as likely as overall global
population growth.

Global population projections continue, therefore, to be based on the assumptions of
the DTM and the main uncertainties relate to making predictions about future fertility
decline in countries that have not completed the DTM. Yet the projections also illustrate
that the population dynamics do not end at Stage IV of the model, but that uncertainty
about the future of population growth or decline in post-transition societies remains,
and moreover that the magnitude of these trends is considerable.

The likelihood of populations decreasing, rather than increasing, is even more important if we focus on subnational areas. One of the most striking outcomes of the global
recession that started in 2008 with the collapse of the American housing market (caused
by the complex trading of sub-prime mortgages) was the demographic collapse of
northern industrial American cities. For example, the population of the city of Detroit
stood at about 2 million people in the 1950s associated with the industrial expansion
around the automotive industry during the first half of the twentieth century. The 2010
US Census reported a population of 714,000 – a fall of 25 per cent over the previous
decade. The population decline was brought about by the collapse of Detroit's industrial
base, but in turn was exacerbated by the problems facing the city that stemmed from
the global recession.

Using a smaller scale also poses more problems for projections, such as the need to
include the dynamics of migration to allow for people moving between regions. The
dynamics of sub-regions cannot therefore be reduced to differentials in mortality and
fertility, which dominate the DTM, as individual mobility increasingly needs to be taken
into account. Of course, the same observation applies for national populations, as the
magnitude of international migration increases.

We shall return to consider the population dynamics of sub-regions and the magnitude of migratory moves in Chapters 4 and 6. In the next sections, we consider how
demographic data is collected and made available.

# Resourcing population analysis

Information on population has been regarded as important for at least as long as the concept of the nation state has existed. Counts of population were carried out in Babylonia around 3800 BC, in China about 3000 BC and in Egypt around 2500 BC. The Bible contains evidence for counts, most famously the census which involved Joseph returning to his home town with his pregnant wife Mary. Whatever one's religious beliefs, people interested in population geography, demography and the recording of related data can view this as evidence that a census was taken and that is it most likely that somebody called Jesus was born at that time in Bethlehem. The information from these early censuses was used to inform rulers as to the extent of revenues they could expect to raise through taxes and to determine the number of people who could be mustered for war.

The strategic importance of population statistics remains today. Many countries operate a national statistical agency to collect and analyse population statistics for a multitude of purposes. For example, in the UK, population statistics directly inform the distribution of funds from central government to the various local councils around the UK. Allocation is based not only on population size within an area but also on other factors such as the population composition and the levels of deprivation that impinge on the costs of local service provision.

From a political perspective, population statistics are regularly employed in the arguments put forward by political parties in support of particular policy programmes. The importance of transparent and responsible use of such statistics was recognised in the UK Statistics and Registration Service Act 2007 which sought to improve declining public trust in official population statistics. This act made the Office for National Statistics independent from government control and created the UK Statistics Authority which oversees the Office for National Statistics and monitors the use of official statistics in the UK.

At a finer administrative level, local councils in the UK usually have a research team who are responsible for the analysis of population statistics to inform and evaluate local policy decisions and resource allocation. Outside the public sector, businesses regularly use data on population to inform and coordinate the marketing and sale of their products in order that their profits are maximised. Marketing companies such as CACI have responded to this demand by developing techniques to classify areas according to their population characteristics and selling this information to businesses, government and academic researchers. For many businesses, the information they collect about their customers (through loyalty cards, delivery information or store credit cards) is increasingly seen as a product in its own right.

Population research needs data and this information is collected in a number of ways. Data to measure population size, characteristics and change can be obtained through the taking of censuses, registration systems, the collection of surveys or the use of administrative data collected for other purposes. The most fundamental information required is the number of people who live in a particular area. But we are usually also interested in how many males and females there are at each age. To calculate population change, we need counts of persons who come into the world (fertility/births data), how many leave us (deaths/mortality data) and how many people move between areas (migration) which may be within a country (sub-national or internal migration) or between countries (immigration and emigration). In many applications we want to know more about people's attributes; their marital status, ethnicity or race, level of educational achievement, employment status, etc. These kinds of data can be used to paint a very detailed picture of the human geographical landscape and to measure the processes of population change.

## Census data

Population statistics are a wide-ranging collection of counts of the people living within a country. The primary source of population data is a census. A census can be defined as the 'total process of collecting, compiling, evaluating, analysing and publishing demographic, economic and social data pertaining, at a specified time, to all persons in a country, or a well delimited part of the country' (UN 1998). The intention of a census then is to collect data about 100 per cent of the target population. Information is usually obtained through a questionnaire 'enumeration' form which contains a mix of tick-box answers and write-in sections.

As noted above, censuses date back to ancient times. The word census is from the Latin verb *censere* meaning to 'estimate' or to 'assess'. The US began taking regular censuses in 1790 with other 'developed' countries soon doing the same. In Great Britain, a population census has been conducted every ten years since 1801 (and is therefore often referred to as the 'decennial' census). There have been two exceptions to this; there was no census taken in 1941 during the Second World War and there was an additional census in 1966 which comprised a 10 per cent sample of the population. The first national census in Australia was in 1911 and since 1961 censuses have been taken in Australia every five years.

The early censuses in England and Wales were, in comparison with recent censuses, rather simple inquiries. The first four censuses in England and Wales (1801, 1811, 1821 and 1831) were undertaken by the 'Overseers of the Poor' in each parish and the information collected included:

- number of houses inhabited, the number of families occupying them and the number of uninhabited houses;

- number of persons and their sex, including men in the armed forces and at sea;

- number of persons engaged in agriculture, trade, manufacturing or handicrafts and other occupations.

In 1841, the duty of filling in the enumeration form was delegated by the Registrar to the 'head of the household' (assumed to be male) changing it to a self-completion questionnaire. Additionally the filling in of all the forms was conducted on the same day across the whole country. The collection of information in this way has remained unchanged since 1841, although the head of household who fills in the form is no longer assumed to be male. Now, the person who fills in the form has to declare how many people usually reside at that address and how many were absent and how many visitors there were on census night. Then the form is returned either online or by post.

Collection of comprehensive data from every person in the country is an expensive, time-consuming and complicated task. Censuses are funded by a country's government and the execution of a census is made the responsibility of an official organisation or department. The census relates to a well-defined area at a specified time and is thus regarded as a population cross-section or 'snapshot'. For accuracy, information must be provided by every person in the country. To approach a 100 per cent enumeration it is not sufficient to rely on voluntary cooperation and so powers of compulsion have to be given to the census-taking organisation. It is usual for these powers to be contained in a legislative act which authorises the taking of a census. Most countries restrict the release of information collected in a census with the confidentiality of information on individuals of paramount importance. The 1920 Census Act in the UK keeps data about specific individuals collected in any census confidential for the next 100 years after each census. The censuses up to 1911 are an invaluable resource for genealogical research and in due course, data from the 1921 Census will become available.

The UN (2011) publishes a useful summary of census taking around the world with dates and copies of questionnaires from a very large number of countries. In the recent round of international census taking, censuses were taken in the US in 2010 and in the UK and Australia in 2011.

It is important to note that a number of countries have taken a decision to collect less information from population censuses. For example, both Canada and the US dropped the full schedule of questions from their most recent censuses, the completion of which was not mandatory in Canada. In the UK, at the time of writing, the Office for National Statistics (ONS) is exploring alternative approaches to collecting data on population counts and characteristics since it is possible that the 2011 Census will be the last census in the UK.

## Population register data

Population censuses are not the only means of collecting information on populations. Some countries, such as the Netherlands and Scandinavian countries, maintain a continuous population register. For example, in the Netherlands, any person who is staying in the country for more than four months is required to register their details in the database of the municipality in which they are living. They are required to provide information on characteristics such as nationality, partners and children and to inform the authorities if they move within the Netherlands or abroad. The essential feature of a population register is that the demographic events, births, marriages, deaths and migration movements, are recorded continuously throughout the year. There is no such register in the UK. Many people would argue that a national register would improve the quality of the UK's population statistics. However, the existence of compulsory identity cards and/or a central dossier is regarded by others as an invasion of civil liberties.

## Vital statistics data

The registration of 'vital' events began in Europe as the ecclesiastical function of recording baptisms, weddings and funerals. In England and Wales, this began during the reign of Henry VIII who ordered parishes to maintain registers with the duty to carry this out alternating between the clergy and laymen. In England and Wales, two Acts of Parliament were passed in 1836, the Births and Deaths Registration Act and the Marriage Act, which set up the system of civil registration. From 1 July 1837, registration of all births, marriages and deaths in England and Wales became the responsibility of the state. The Acts also established the role of 'registrar general' (the country's official population statistician) the first of whom was William Farr. He set up a system to routinely record people's cause of death so that mortality rates of different occupations could be compared (Halliday, 2000). After an Act of Parliament in 1874 it became compulsory for all births and deaths to be registered. In the US, although the collection of vital statistics was carried out in connection with the 1850 Census and through registration systems in some states, there was no country wide framework for the regular collection of births and deaths data until 1933.

Currently in England and Wales there is a legal requirement to register a birth within 42 days and a death, normally within five days. As a data source, the vital statistics on births and deaths are considered to be a high quality resource. Births and deaths data are used to compute fertility and mortality rates which inform population geographers on how natural change (the difference between levels of births and deaths) is affecting population change. Data from both parish records and vital statistics provide an important resource for historical demographers.

## Sample survey data

Large scale surveys can be carried out to inform on many more aspects than can be included on a census questionnaire. Surveys may be carried out on the instigation of government departments or as part of an academic project, for example. Rather than self-completion questionnaires, these surveys are usually carried out by teams of trained interviewers and can comprise a large number of questions which can cover a range of information about a person and their household as well as their views on certain topics such as government policy or about the area in which they live. A census aims to capture information about the whole population. A survey of the kind we are referring to here aims to interview thousands of people who form a representative sample of the national population or of a particular target group (the elderly, for example). Though surveys do not provide comprehensive population coverage, they enable, at a reasonable cost, more detailed information to be collected than is the case in a census.

In the UK, a number of large scale surveys are carried out that are explicitly focused on particular population or household aspects and include the Labour Force Survey and the Health Survey for England amongst many others. The surveys may have full coverage for the UK or may be confined to one of the constituent countries and tend to focus on the household population excluding institutional populations (e.g. prison, army or care home populations). There are a large number of surveys which have great potential to inform population and social geographers about population characteristics. For example, the Health Survey for England (and equivalents in other parts of the UK) has a range of questions on health as well as 'objective' measures such as records of blood pressure and body mass index. The Labour Force Survey provides information on the UK labour market and also includes a longitudinal element where the characteristics of individuals are recorded over time. The survey has been used for many research purposes including, for example, to estimate fertility rates by ethnic group (Norman et al., 2010; Dubuc, 2009). The UK Data Service archive provides support for these survey datasets.

Across Europe, there are international surveys which include the European Social Surveys and Eurobarometer Survey Series. The European Social Survey is designed to capture how beliefs, attitudes and behaviours are changing throughout Europe. In 2011, the survey was preparing for its sixth round and covered more than 30 countries. The Eurobarometer Survey is a repeat cross-sectional survey, that has been funded through the European Commission's Framework Programmes, the European Science Foundation and national funding bodies in each country and is designed to monitor social and political attitudes. In the US, the American Community Survey is taken to complement the census and the US National Library of Medicine makes a large number of surveys available. In developing countries, the Demographic and Health Survey (DHS) has been carried out since 1984 in over 84 countries and provides a very informative international source. The DHS is funded by the US Agency for International Development (USAID) with contributions from other donors including UNICEF, UNFPA, WHO, and UNAIDS. The aim of the DHS is to advance global understanding of health (e.g. malaria, HIV) and population trends (e.g. levels of fertility and attitudes to contraception).

## Administrative data

The sources referred to above such as censuses and vital statistics data are collected to inform national and local governance. However, data collected during administrative processes can be used for the analysis of population and society even though they were not originally collected for this purpose. Particularly informative sources can be social security/benefits claimant data on unemployment and sickness benefits, council tax records, counts of the electorate and medical records.

## Qualitative data

The study of populations has mostly utilised statistical or quantitative data. However, in recent years, population researchers have sought to extend their analysis of demographic processes through the use of mixed-method research which combines quantitative data with qualitative data. For example Bob Woods (2006) uses poems and pictures of children throughout history, and particularly accounts of child death, to refute the hypothesis of parental indifference to infant death in historical populations. Migration researchers have also sought to combine quantitative analysis with the use of biographical data to enrich accounts of mobility and explore the variation of experiences and the importance of cultural determinants (Halfacree and Boyle, 1993; Ní Laoire, 2000).

# Key methodological approaches for population research

What you do with population data largely depends on the questions you are interested in addressing and the specification of the data to which you have access. Here we outline some of the distinctions in style of analysis which, in the main, depend on data type.

## Geographical or individual level data

Whilst all data are originally derived from individuals, some are available to practitioners for geographical areas and some as individual records. Much of the census, vital statistics and administrative data noted above will be aggregated from the original raw form (a person's census questionnaire or a birth certificate, for example) into data for areas. Large scale survey sample data are usually available as individual records, presented anonymously so that individuals cannot be identified. It is important to know whether your dataset is inherently geographical or individual because, although the research questions you are posing may be very similar, the methods you use for analysis are likely to be somewhat different.

   The geographies for which data may be available largely relate to the application for which the data were collected. Since the need for good governance has often driven the process, census, vital statistics and administrative datasets are usually available for national and local government areas and for electoral geographies. Part of the motivation of census taking is to make data available for very small areas so geographies are defined for the release of census data with populations or household counts just large enough so that confidentiality of individuals is maintained. There tends to be greater sociodemographic detail available for larger areas than for smaller ones. So, for example, a census table for England may have employment status available by sex and age in five-year groups but at a more local level there might only be a breakdown by sex. Data tend to be real number integers and for a simple analysis allow percentages to be calculated (e.g. males who are unemployed, divided by the total of males who are economically active, multiplied by 100).

Individual level data from a large scale survey tend to be much more flexible than area level data in terms of the variable definitions and cross-classifications that can be achieved. For instance, age may be recorded by single year allowing re-classification into groupings relevant for a particular research question. Whilst survey data may have very detailed sociodemographic information, it is rare that small area geographical locations are provided.

Another important difference between the census and individual level survey data is that census data includes (almost) all the population whereas survey data are invariably a sample. This may mean that different analysis methods are applicable, for example surveys often include survey weights that take into account disproportionate sampling or non-response of a particular population group. Unlike census estimates, survey estimates are usually reported with a confidence interval. This confidence interval provides a measure of the uncertainty associated with a sample estimate reflecting the fact that were a slightly different sample collected a different estimate would result.

## Time frames

Many datasets are referred to by name and with a year. So, taking data available in the UK as examples, we talk of the 2001 Census, the vital statistics for 2004 or the Health Survey for England (HSE) 2008. The census is taken on one day and is regarded as a snapshot in time (though people may not fill in their census forms on the specific day!). The vital statistics are collated over a calendar year (all births and deaths occurring between 1 January and 31 December). The large scale surveys like the Health Survey for England are often carried out over the course of several months whilst a team of interviewers carries out the field work.

A dataset (comprising individual or area information) at a particular point in time is often referred to as containing cross-sectional data. A series of cross-sections creates a *time-series* of data. A time-series of decennial census or annual vital statistics or survey data can be used to identify trends in a particular aspect or population characteristic. The overall time frame of a study is potentially only limited by the availability of data.

*Longitudinal* studies track the same individuals over time and these enable a *life course* perspective to be taken. These studies may interview people at annual or more periodic intervals and may be constructed as a *cohort* study (about individuals all born within the same time period) or as a *panel* study (about a representative population group). In addition to interview approaches, some studies may capture data on individuals from censuses and link a person's records to administrative datasets to build up a statistical picture of people's lives. These styles of data have restricted access due to the level of personal information included and are complex to analyse. The results of such studies, though, can be extremely informative. Data sources of note in the UK include the National Child Development Study (a cohort born during 1958), the British Household Panel Survey and its successor Understanding Society, the Life Opportunities Survey and the Longitudinal Studies (of England and Wales and separately of Scotland and of Northern Ireland) which use census records on a sample of individuals. Examples of birth cohort studies outside the UK include those carried out in Denmark, France, the US and Quebec. There are a number of longitudinal studies which focus on the elderly such as the English Longitudinal Study of Ageing and similar studies in the US, Australia, Canada, Japan and the Netherlands.

As discussed in the previous chapter, the concept of a *transition* is a useful device to analyse change. The time scales can vary but of particular interest in geographical studies are the commuting movements from home to work (on the same day) or migration moves from origin to destination (whether there has been a change of address from one year to the next). Similarly, individual level cohort studies may identify transitions in and out of employment or ill-health in annual or decennial time increments. Social mobility may be defined in a variety of ways but involves a transition from one state to another (for example, manual to non-manual employment).

## Summary

This chapter has considered the importance of population projections and the need to interpret these carefully, and the sources of data for population analysis. The chapter can be summarised as follows:

- Global population projections draw upon the concepts of demographic change outlined in the DTM. As we move to finer geographical scales the DTM becomes less relevant and the uncertainty associated with population projections increases. A range of people including academics as well as politicians and planners in the public and private sectors have an interest in quantifying and understanding the nature of populations at points in time and their change over the course of time. For example, national and local government officers have a need to have counts of persons by age-sex and area in order to provide appropriate numbers of schools, houses, care homes and staff.

- To aid ongoing planning, projections of future population size and age structure are needed. Similarly business applications also need this type of information, again, so that appropriate goods and services can be most efficiently supplied. From a political perspective, population analysis is essential to inform debates on topics such as overpopulation, population ageing and migration. A wide variety of data exists to underpin studies with methods of analysis methods depending on data types and the questions being posed. This is a data rich subject area with census, administrative and survey data available in most countries across the world. Sometimes, researchers will find themselves rather overwhelmed by data in terms of both the quantity and complexity of information. At other times, data restrictions may leave the analyst somewhat frustrated.

# Recommended reading

## Data sources

Centre for Longitudinal Studies: www.cls.ioe.ac.uk/
Demographic and Health Surveys: www.measuredhs.com/aboutdhs/
UK Data Service
UK Office for National Statistics: http://www.statistics.gov.uk/hub/population/index.html
United National (2010) World Population Prospects: http://esa.un.org/wpp/
US Census Bureau: www.census.gov/
US National Library of Medicine: www.nlm.nih.gov/hsrinfo/datasites.html

## Methods resources

The following books provide detailed information and instructions about demographic data and methods:

Dale, A., Fieldhouse, E. and Holdsworth, C. (2000) *Analysing Census Microdata*. Arnold: London.
Diamond, I. and Jefferies, J. (2001) *Beginning Statistics: An Introduction for Social Scientists*. London: Sage.
Fotheringham, A.S., Brunsdon, C. and Charlton, M. (2005) *Quantitative Geography: Perspectives on Spatial Data Analysis*. London: Sage.
Lloyd, C.D. (2010) *Spatial Data Analysis: An Introduction for GIS Users*. Oxford: Oxford University Press.
Marsh, C. and Elliott, J. (2008) *Exploring Data*. Cambridge: Polity Press.
Newell, C. (1988) *Methods and Models in Demography*. Chichester: Wiley.
Rogerson, P.A. (2006) *Statistical Methods for Geography*. London: Sage.
Rowland, D.T. (2003) *Demographic Methods and Concepts*. Oxford: Oxford University Press.

# 4

# POPULATION STRUCTURES

This chapter investigates the age-sex structure of populations. As we will see, there can be substantial variations in populations by the number or proportion of people in each age-sex group. Knowledge of population size and structure is fundamental to both research and applied work about the population itself and because the population structure feeds into our understanding of demographic rates of fertility, migration and mortality. In this chapter we explore:

- why it is important for practitioners to know about population structure;

- how population structure can be illustrated using population pyramids and measured using dependency ratios;

- variations in population structures between countries over time;

- variations in structures for subnational areas and for different ethnic groups;

- the inter-relationship between population structure and demographic events;

- how policy may directly or inadvertently affect population structure.

## Why and how do we investigate population structure?

The provision of goods and services is often age and/or sex specific. For educational applications, the number of children of school age determines the current demand for school places. The number of very elderly creates demand for medical services and care home places. Business applications need to know the size and characteristics of populations for marketing purposes. Direct intervention by government can seek to influence population size and structure to achieve certain goals but whether policy works or is advisable is subject to wide debate. After the count of an area's total population, the most fundamental information we therefore need to know about is the age-sex structure of the population; i.e. how many or what proportion of the area's population are male or female and are of a particular age-group. Unless there is a good understanding of the population size and structure then any study of population whether academic and interpretive or applied and practical might be ill-informed.

Counts of the population by age and sex are used in demographic measures such as fertility (birth), migration (residential moves) and mortality (death) rates. For the latter in particular, when you are comparing two or more populations unless you account for structure, the result may be biased. This is because the risk of mortality varies with age so if an area has a young population structure we would not expect so many deaths during a given year than in a population with an older age structure. This is explored further in Chapters 5 and 9. Since many migrants, people who change their residential address, are young adults, a population with a young age structure is likely to be more mobile. This is explored further in Chapter 6.

Commonly used devices to illustrate and measure population structure are population pyramids and dependency ratios. These measures are detailed in Box 4.1. The examples in this chapter will use these devices to compare population structures of different countries and to look into variations within the UK by both geographical areas and by population subgroup. Data for international examples have been obtained from the UN Population Prospects Database. Subnational and population subgroup examples in the UK are underpinned by data from the 2001 Census. During this chapter we use examples from the early twenty-first century and explore the recent past to see how contemporary population structures have developed as well as considering possible futures through the use of population projections. We conclude the chapter with an overview of population structures in policy situations.

---

## Box 4.1  Illustrating and measuring population structure

All data here are for the UK in 2005 and have been obtained from the United Nations Population Division (2009) *World Population Prospects. The 2008 Revision*

### Population pyramids

A population pyramid is a very useful device through which to illustrate age structure. The bars in the pyramid represent an age-group with males to the left of the centre of the 'x' axis and females to the right. The length of the bar can represent counts or percentages. Here they show each age-sex group as a percentage of the total population. The youngest age-groups are at the bottom of the 'y' axis on the graph and the oldest age-groups at the top. As with any graphs, the scale used can

*(Continued)*

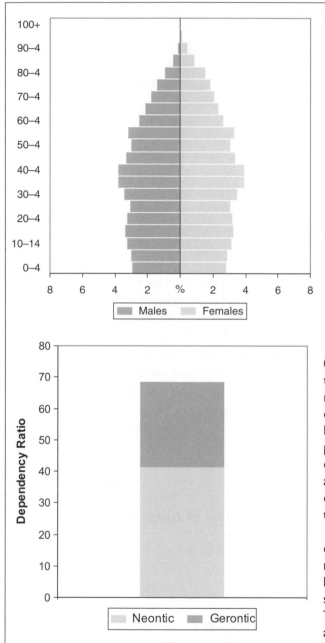

Males    Females

Neontic    Gerontic

distort the picture portrayed. In this chapter, consistent scales are used wherever possible.

## Dependency ratios

Dependency ratios are a relatively simple calculation. The implication of the word 'dependency' in this setting is that some of the population are dependent on others; usually in the sense of whether or not people are economically active. The conventional assumption is that the young and the elderly are dependent on those of working age. Here we have calculated the neontic/ young dependency ratio as the count of the population aged 0–19 divided by the count of the population aged 20–64 and multiplied by 100. The gerontic/ elderly dependency ratio has been calculated as the count of persons aged 65 and over divided by those aged 20–64 and multiplied by 100. The total dependency ratio is the sum of these.

The age-groups used for the dependency ratio calculations may vary, so comparisons between pre-calculated data sources may be problematical. There is debate on where the age boundaries should be, given different ages at which young people leave education and become economically active and variation in retirement age when people leave the workforce and thereby become economically inactive. Indeed not all persons of working ages are economically active to be able to support the dependent population.

# National variations in population structures

A classic use of population pyramids is to illustrate a country's stage of development since distinctive pyramid shapes can be observed for specific combinations of mortality, fertility and migration rates. There are links here to the Demographic Transition Model (DTM) which we explored in Chapter 2. For 2005, in Figure 4.1a the population of Zimbabwe is shown to have very high birth rates and good survival in early years but from late teenage years the size of the bars reduces rapidly indicating high death rates with each successively older age group. The small bars at older ages indicate short life expectancy. The convex shape of the pyramid from age 20–24 is typical of Stage I of the DTM but not the relatively low child mortality suggested by Zimbabwe's pyramid. Figure 4.1b illustrates the population of India in 2005. This shape of pyramid is typical of Stage II/III of the DTM. Whilst there are still high birth rates, survival to the next age-group throughout the whole population is improved resulting in a more 'middle-aged' population with overall somewhat longer life expectancy. With these pyramids, the populations in both Zimbabwe and India are likely to be expanding in size.

Stages III and IV of the Demographic Transition Model would see a pyramid with convex sides as death rates fall and life expectancy increases. Low birth rates in Stage IV would 'undercut' the pyramid in the youngest age groups. For 2005 in Sweden (Figure 4.1c), the population is probably beyond Stage IV with a pyramid having high survival to the next oldest group. This is even more exaggerated in Japan which exhibits extremely low birth rates. In both countries, there is long life expectancy. Whilst the population in Sweden might be considered to be stable, over time the national populations of both Sweden and Japan are likely to contract without immigration of persons from other countries.

Figure 4.2 (page 55) illustrates the dependency ratios for the same countries in 2005. Since the total dependency is almost 140 in Zimbabwe, this means that more people are dependent on the working-age population than are in that broad age-group. Given the high numbers of young and few elderly shown in Figure 4.1a it should be no surprise that this large dependency is predominantly neontic (i.e. youthful, see Box 4.1) whereby there are 129 persons aged less than 20 per 100 persons between the ages of 20 and 65. Although life expectancy in India is somewhat longer than in Zimbabwe the gerontic (i.e. elderly, see Box 4.1) dependency ratio is the same. The neontic ratio is much lower though partly due to the somewhat lower fertility rates but largely because of the much better survival of adults through the working ages. Overall dependency ratios in Sweden and Japan are much lower largely because of the low birth rates and relatively small populations aged 0–19. The longer life expectancies result in higher gerontic dependency ratios, particularly in Japan where it is larger than the neontic ratio.

Population ageing involves a redistribution over time of an area's population towards older ages and is evidenced by an increasing proportion of the population who are elderly. This is not just driven by increasing life expectancies leading to more people

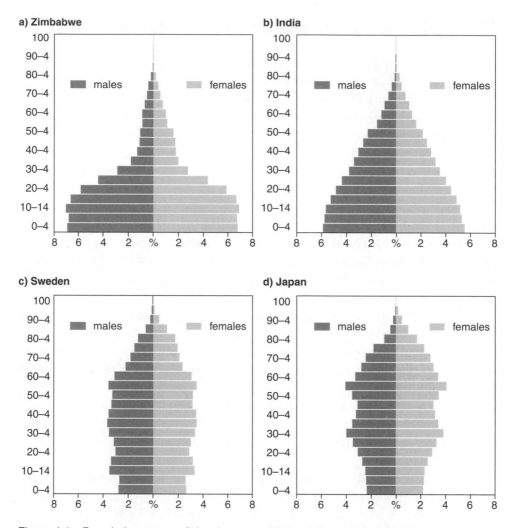

Figure 4.1    Population stage of development: National Populations, 2005

*Source:* UN Population Division (2009b) *World Population Prospects. The 2008 Revision*

living into older ages but also to decreasing birth rates which leads to fewer children and then later to fewer working-age adults. Since fertility is declining rapidly in many developing countries, so too will their populations be ageing. As survival into older ages improves, the rate of ageing will increase.

## Population structure as evidence of the past and predictor of the future

The population structures illustrated above have largely been described in a cross-sectional way as a snapshot of the population at a given time point. Hinde (1998) usefully points out

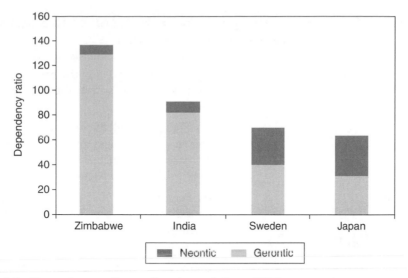

Figure 4.2    Dependency ratios: national populations, 2005

*Source:* UN Population Division (2009b) *World Population Prospects. The 2008 Revision*

that the age-sex structure of an area is influenced by demographic events in the past so that a pyramid is effectively a historical record of *past* population. Hinde (1998: 163–4) provides a detailed interpretation of the age-sex structure of France in 1992.

The population at any time point is a function of the previous population plus births which have occurred and minus deaths. Also, the population size and structure will be affected by migration in and out of an area. In Figure 4.3a, a population pyramid of the UK population in 2001 is annotated for several features. Those aged 0–4 in the population (a) will have been born within the previous five years whereas those aged 85 or over (b) were born before 1916. At ages over 70 there is an increasing predominance of females due to their longer life expectancy. Any of these people may have been born in this area and aged *in situ* or have moved from elsewhere. Note that the smaller the geographical area, the more likely it is that migration will have affected population structure and the greater the size of the impact.

Previous rises and falls in fertility, and the number of live births, are evident. In Figure 4.3a, those aged 50–54 (c) were born in the period following the Second World War when troops were returning home. From the late 1950s into the 1960s there was a 'baby boom'. In 2001 these persons were in their thirties and forties (d) and are represented by the swell in the bars at these ages. Following the baby boom, there was a 'baby bust' when fertility rates fell rapidly and remained low during the 1970s accounting for the pinch in the pyramid (e). The slight increase around the 10–14 age-group (f) is a 'baby boom echo' as the relatively large baby boom generation had children themselves. Even though we have focused on births here, the size of any cohort (a group of people born in the same time period) will also be affected by migration and mortality.

As we have seen, a population pyramid is evidence of the demographic history of an area, but to a large degree, an area's age-sex structure is highly predictive of the *future*

## Figure 4.3　Evidence of the past and predictor of the future: age-sex structure, UK, 2001

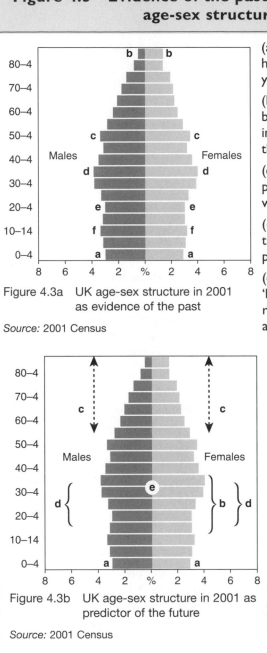

Figure 4.3a　UK age-sex structure in 2001 as evidence of the past

*Source:* 2001 Census

Figure 4.3b　UK age-sex structure in 2001 as predictor of the future

*Source:* 2001 Census

(a) Persons aged 0–4 in the population will have been born within the previous five years.

(b) Persons aged 85 or over were born before 1916. At ages over 70 there is an increasing predominance of females due to their longer life expectancy.

(c) Persons aged 50–54 were born in the period following the Second World War when troops were returning home.

(d) From the late 1950s into the 1960s there was a 'baby boom'. In 2001 these persons were in their thirties and forties.

(e) Following the baby boom, there was a 'baby bust' when fertility rates fell rapidly and remained low during the 1970s accounting for the pinch in the pyramid.

(f) The slight increase around the 10–14 age-group is a 'baby boom echo' as the relatively large baby boom generation had children themselves.

(a) Persons aged 0–4 in 2001 will be five years older in 2006, as will each cohort presuming that all persons survive.

(b) The number of women of child-bearing ages (approximately age 15–44) will largely determine the number of births (those who will be aged 0–4 in 2006).

(c) The risk of death increases with age so there are more likely to be losses from the cohorts for older ages.

(d) Migration tends to be concentrated in young adult ages, so the size of the cohorts will be affected by the impact of migration; specifically whether or not more people move into, or away from the area.

(e) When the baby boom generation reach retirement age, the UK will have a 'superageing' society.

population. We again use the structure of the UK population in 2001 as an example. Since this pyramid has five year age-groups it is logical to describe time into the future in five-year increments. In Figure 4.3b, those aged 0–4 in 2001 are annotated (a). In 2006 these people will be aged 5–9, five years older as will each cohort presuming that all persons survive or do not move out of the country during 2001–2006. The size of the cohort might be increased as more immigrants move into the UK than the number of emigrants who move to another country.

The demographic events of births to women, deaths and migration have very distinctive age profiles and the age-sex composition of an area is highly predictive of the number of events which will occur. So, in Figure 4.3b, the number of women of child-bearing ages (approximately age 15–44) (b) will largely determine the number of births (those who will be aged 0–4 in 2006). The risk of death increases with age so there are more likely to be losses from the cohorts for older ages (c). Migration tends to be concentrated in young adult ages (d) and the size of the cohorts will be affected by the impact of migration; specifically whether or not more people move into, or away from the area. When the baby boom generation (e) reach retirement age, the UK will have a 'superageing' society.

## The development of contemporary age-structure

So if a pyramid at a point in time represents a historical record, then we can add to the record by considering previous population structures as evidence of how the composition developed. First, we will consider the international pyramid examples used previously in Figure 4.1 and then look at changes in dependency ratios between 1950 and 2005.

Using the UN data, Figure 4.4 shows that in 1950 (the lighter grey bars of the pyramids), the population in Zimbabwe was very youthful with high birth rates and high mortality with each successive cohort such that the shape of the pyramid is concave. In India, although survival from the population aged 0–4 is not high, mortality is lower in India compared with Zimbabwe, and the shape of India's pyramid in 1950 is steep but not concave. Hence, life expectancy is somewhat longer. In Sweden, birth rates in the 10 years prior to 1950 were relatively high (post war spikes), compared with the next older cohorts (suppressed to a degree during the Second World War). The pyramid is then parallel sided up to age 40–44 with longer life expectancy in older ages. Japan's population pyramid resembles that of India in 1950 although there is something of a gash for males aged in their twenties and thirties probably due to deaths during the Second World War.

A time-series of pyramids is somewhat cumbersome, but line graphs of dependency ratios over time can be revealing. Figure 4.5a (on page 59) shows that the neontic ratios in Zimbabwe and India are higher than the other countries throughout the 1950–2005 period. In both of these countries, there is a rise in young dependency throughout the 1950s and 1960s and continuing into the 1980s in Zimbabwe as the population aged

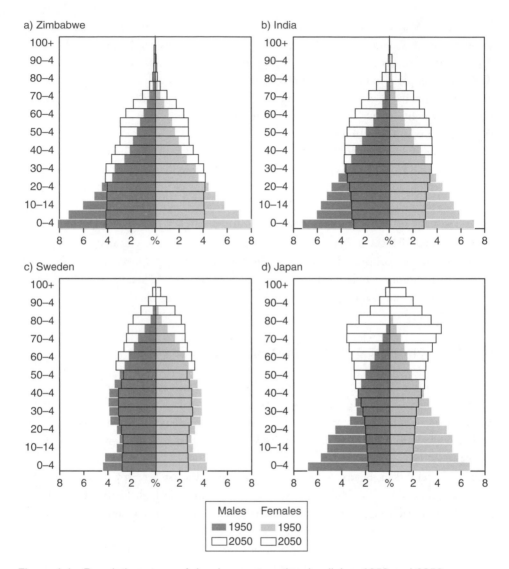

Figure 4.4　Population stage of development: national policies, 1950 and 2050

*Source:* UN Population Division (2009b) *World Population Prospects. The 2008 Revision*

0–19 grows relative to the size of the working aged populations. In both countries the ratios fell to 2005. Whilst the neontic ratio in Japan was almost as high as the ratio in India in 1950 this measure falls rapidly all through the period to be the lowest of these countries in 2005. In contrast, the neontic ratio in Sweden was relatively low in 1950 and stays remarkably steady only reducing marginally by 2005.

As might be expected for countries with relatively short life expectancy, both Zimbabwe and India have low gerontic dependency ratios (Figure 4.5b). The numbers of elderly only increase marginally between 1950 and 2005 and there is scarcely any

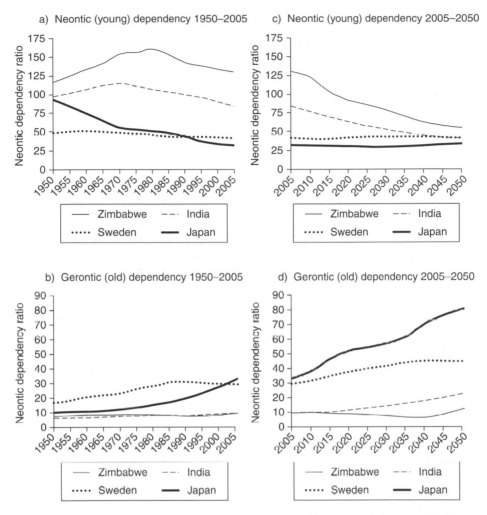

Figure 4.5  Neontic and gerontic dependency ratios: national populations, 1950–2005 and 2005–2050

*Source:* UN Population Division (2009b) *World Population Prospects. The 2008 Revision*

change in the ratios over the time period. In 1950, Sweden had a relatively high geron-tic dependency ratio and this increases steadily into the 1980s before reducing marginally by 2005. Whilst Japan's gerontic ratio was only a little above those in India and Zimbabwe in 1950, from the 1970s onwards the population in Japan has been ageing rapidly suggesting improving life expectancy.

## Population age structure in the future

A regularly used demographic method through which to project the age-sex counts of population in the future is a 'cohort component' model. In the calculations, this model

takes a 'base' population age-sex structure and ages people on by the time period incre-
ment, adds in the births likely to have occurred and subtracts the deaths. Allowances are
made for migration, movements of people in and out of the area. Many official, academic and business personnel carry out population projections since knowledge about
how population size and structure will change in the future is important in a wide
variety of applications including the likely demand for housing, schooling, employment,
services and pensions as well as for business marketing.

The outline bars in Figure 4.4 illustrate the UN projections of the populations in
2050 which were previously described for 1950 (the grey bars) and for 2005 in
Figure 4.1. Numbers of births in Zimbabwe (Figure 4.4a) are projected to have fallen
resulting in a narrower base to the pyramid. Mortality has also reduced with more
people surviving from one cohort to the next oldest and with more people surviving
to middle and older ages. India's population (Figure 4.4b) has moved to being a 'stable'
shape resembling that of Sweden in 2005 (Figure 4.1c). Numbers of births have
become low, survival to middle age is high and the population overall has aged.
Though some ageing is evident compared with 1950, Sweden's population structure (Figure 4.4c) has not changed much between 2005 and 2050 justifying the
description of being 'stable' during this period. The population ageing processes carried forward in the cohort component model are clear with the swell in the 2005
pyramid for the age-groups 10–14 and 15–19 (Figure 4.1c) remaining large cohorts
when aged 55–59 and 60–64 in 2050. Similarly in Japan, the cohort aged 30–34 in
2005 (Figure 4.1d) has aged on to be aged 75–79 in 2050 (Figure 4.10d) to be part of
a superageing society with a predominance of elderly and a steeply undercut pyramid
as each successive five-year birth cohort is smaller than the one for the previous five
years. The shape of Japan's pyramid projected for 2050 is almost upside down compared with the situation in 1950; a remarkable change in 100 years.

Along with reduced birth rates, the improved survival into adult ages results in
the neontic dependency ratio falling steadily in Zimbabwe between 2005 and
2050 (Figure 4.5c). To a lesser extent, the same patterns can be observed for India. The
stability in Sweden's population is evident since the almost unchanging neontic ratios
between 1950 and 2005 (Figure 4.5c) continue up to 2050. The fall in Japan's neontic
ratios between 1950 and 2005 appear to have bottomed out staying almost constant
between 2005 and 2050.

Although there has been increased survival in Zimbabwe (Figure 4.5d) this does
not yet lead to increasing gerontic ratios; largely because it is the working age
population which is increasing in size. In India though, with lengthening life
expectancies, the gerontic ratios increase steadily to 2005. The slight reductions in
gerontic ratios recorded for Sweden in the late twentieth century (Figure 4.5b) turn
around with a gradual increase projected to 2050. The rises in gerontic ratios in
Japan observed from the 1980s onwards (Figure 4.5b) continue at a strong pace up
to 2050 (Figure 4.15d).

# Subnational variations in population structures

Since different types of people live in different places we should explore whether population structure varies at subnational levels: areas within a country. In this case, we investigate age-sex composition for areas within the UK using data from the 2001 Census, as illustrated in Figure 4.6. When there are a large number of areas within a country, it is common to stratify a phenomenon of interest across a classification of area types. Here we will use a geodemographic style of classification, the 'Supergroups' (Vickers and Rees, 2006). This classification groups areas together which are sociodemographically similar and labels each type with a descriptive name.

Figure 4.6a shows the population structure for the Supergroups in 'London' (combining London Centre, Cosmopolitan and Suburbs). The proportion of children aged 0–4 is in line with the national picture but the reduction in successive ages up to age 20 is unusual. This is likely to indicate migration away from these areas rather than mortality since deaths for children aged 5 to 19 are very rare in the UK. This pyramid for London is dominated by the very large proportions of adults in their twenties and thirties. Above these ages there is a rapid reduction in population such that a concave shape is evident. 'Cities and Services' areas (Figure 4.6b) have a much smoother profile with a relatively large proportion of adults in their twenties and thirties. Both males and females show fair survival into old age.

Somewhat surprising, given their very different names, is that 'Prospering UK' and 'Mining and Manufacturing' areas have very similar age structures (Figure 4.6c and d) with a lack of young adults, a rather 'middle-aged' population including a spike at age 50–54 (more of this later) and high chances of survival into older ages. Whilst the age-sex structures appear similar, the populations are of a very different size. In 2001, nearly 21 million people lived in Prospering UK areas and 12.5 million people in Mining and Manufacturing areas.

'Coastal and Countryside' areas (Figure 4.6e) have very similar proportions of children aged 0–19 as the Prospering UK areas and Mining and Manufacturing areas. There is even more of a waisted shape with a lack of young adults but then these Coastal and Countryside areas are characterised by late-middle aged and elderly populations. Whilst in other area types there is evidence of a predominance of female elderly since male life expectancy is somewhat shorter, in Coastal and Countryside areas this is more marked.

The dependency ratios for Cities and Services, Prospering UK and Mining and Manufacturing are remarkably similar (Figure 4.7). These populations may live in different types of places and have different attributes in other ways (such as employment sector, housing tenure, etc.) but their age-sex structures have only minor differences. The neontic dependency ratio is similar in Coastal and Countryside areas but the gerontic ratio is much larger confirming the observation made of the population pyramid (Figure 4.6e) that this area type is distinctive by its elderly structure.

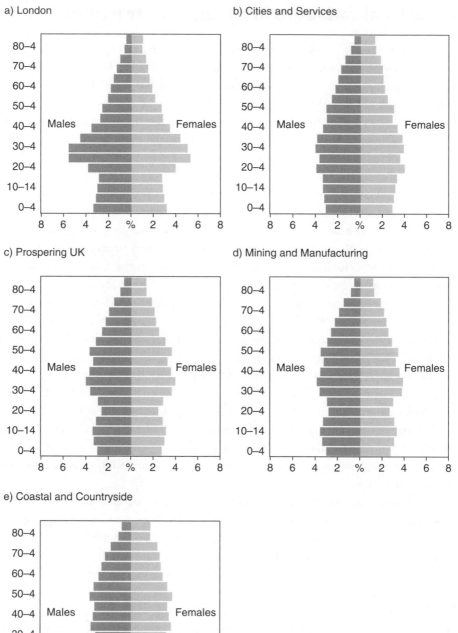

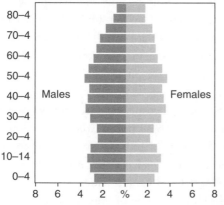

Figure 4.6   Population structures by Supergroup type, UK 2001

*Source:* 2001 Census

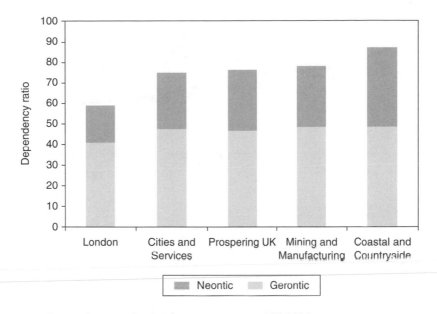

Figure 4.7    Dependency ratios by Supergroup type, UK 2001

*Source:* 2001 Census

Similar to the international examples given above, the development of subnational variations in population structure by area type can be tracked over time in the UK. Dependency
ratios every five years since 1981 up to 2031 have been calculated for the Supergroups and
are illustrated in Figure 4.8a and b. In all areas between 1981 and 2006 the young dependency (Figure 4.8a) falls with the relative position of the areas staying very similar over the
25-year period. Apart from in 1996, London has the lowest neontic ratios, largely because
this area has a large working-age population. There is more change with the gerontic ratios
during this time period (Figure 4.8b) with more difference in the ratios at the end of the
time period than the beginning. This suggests that the area types are not ageing at the same
rate. Both Prospering UK and Mining and Manufacturing area types experienced increasing
gerontic ratios but Cities and Services and especially London have falling ratios. The Coastal
and Countryside areas have the highest gerontic ratios with the level staying consistent over
time. Since the young dependency is falling, these areas are still relatively ageing.

   Between 2006 and 2031 neontic dependency ratios are predicted to decline, but at a
slower pace. Cities and Services areas maintain their neontic ratios more than the other
areas but the biggest reductions are in Mining and Manufacturing areas which by 2031
have the lowest neontic ratios in the UK. The falls in gerontic ratios in London evident
between 1981 and 2006 appear to have bottomed out staying steady up to 2031 (Figure 4.8b).
In the other area types, the gerontic ratios increase steadily between 2006 and 2031 in
Mining and Manufacturing areas but to a greater extent in prospering UK and Coastal
and Countryside areas. In these areas, given the falling neontic ratios, the populations
are projected to age quite rapidly.

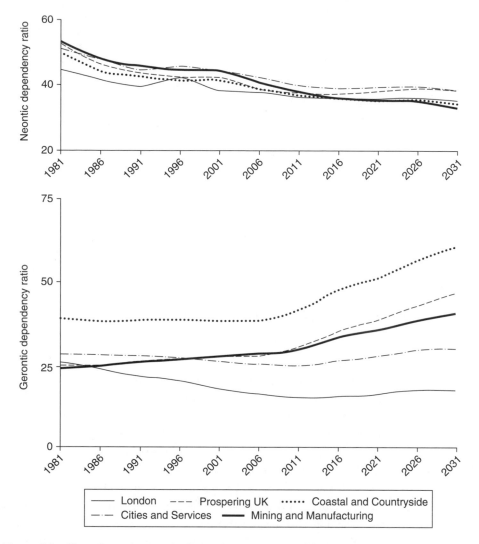

Figure 4.8  Neontic and gerontic dependency ratios by Supergroup type, UK 2001

*Source:* 2001 Census and subnational population projections

## Variations in structure by population subgroup

A census or survey can record a range of attributes about people: their occupation, their country of birth and their ethnicity, for example. In a variety of academic and policy settings, knowledge about population subgroups can be essential for both research and applied reasons. Population subgroups with distinctive age-sex profiles would include residents of communal establishments (hospitals, care homes, boarding schools, armed forces bases, prisons, etc.) as well as people in residential housing. Here as examples we consider the age-sex structure of three ethnic groups in the UK.

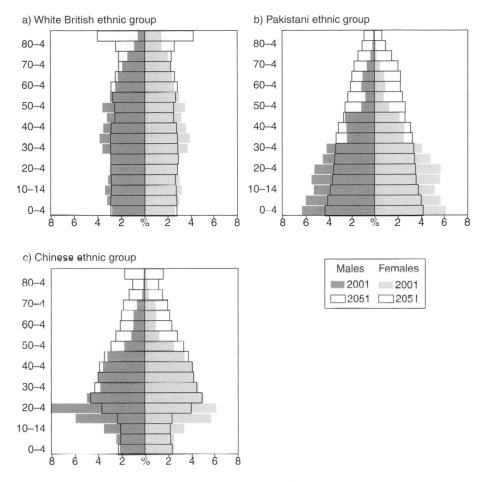

a) White British ethnic group

b) Pakistani ethnic group

c) Chinese ethnic group

| | Males | Females |
|---|---|---|
| 2001 | 2001 |
| 2051 | 2051 |

Figure 4.9    Population subgroups within the UK: ethnic groups in Yorkshire and the Humber 2001 and 2051

*Source:* 2001 Census and ETHPOP database available at: http://ethpop.org/

Figure 4.9a (grey bars) illustrates the structure of the White British ethnic group for 2001 in Yorkshire and the Humber, a region in England. This is very similar to the age-sex structure in Sweden (Figure 4.1c) with a move towards a parallel sided pyramid and a relatively large proportion of elderly. In 2001, the Pakistani ethnic group shows a much younger age structure with substantially larger proportions of young people (Figure 4.9b). The predominance of males aged 40 and over will reflect that these persons were largely immigrants from Pakistan and international flows like this tend to be male dominated. Most of the younger Pakistani ethnic group will have been born in England. The Chinese ethnic group in 2001 (Figure 4.9c) comprises a household residential population which will be a mix of people born in both England and overseas and a large student community who have come to the UK to study. Without the spikes at ages 15–25, the pyramid would be more parallel sided and would reflect a mature population. Chinese birth rates are low resulting in a somewhat undercut pyramid.

Taking these ethnic group populations in 2001 as the base, projections of future sub-national populations to 2051 have recently been published (Norman et al., 2010; Rees et al., 2011). For the Yorkshire and the Humber region of England, Figure 4.9a–c (hollow bars) has the same ethnic groups with their populations projected into the future. By 2051, the White British ethnic group (Figure 4.9a) displays an almost parallel-sided pyramid with the attributes of a stable population. There is projected to be a very large accumulation of elderly persons aged 85 and over with little difference between male and female life expectancies. The Pakistani ethnic group has developed from being a very youthful population with relatively few middle-aged and elderly residents (Figure 4.9b). The pyramid still represents a growing population though with relatively high birth rates and higher chances of survival to the next oldest cohort. Life expectancy for the Pakistani group is projected to improve overall. The low birth rates of the Chinese group evident in 2001 (Figure 4.9c) are projected to persist. It appears that a proportion of the large cohort of a student aged population present in 2001 has stayed in the area. It would be expected that persons attending a higher education establishment will take up postgraduate and employment opportunities and therefore stay in the locality. Since migrants have a distinctive age structure and may differ in their attributes compared with the receiving community, the population composition can be affected in terms of age-sex structure, ethnic mix and ongoing demographic behaviour. However, the degree to which international migrants, whether moving for educational or employment purposes, settle in their country of destination is hard to ascertain.

## Population structure/demographic events interrelationship, population structure and policy

There is a fundamental inter-relationship between the occurrence of demographic events (births, deaths and migration) and a population's age-sex structure. The number of demographic events in the past (whether very recently or over the lifespan of the oldest residents) determines the structure at a point in time. So, we can say that demographic events create the population structure. The population structure though determines the ongoing number of demographic events since these are highly age and sex specific. Thus, we can *also* say that age-sex structure leads to events, namely the number of female births 15 to 49 years ago influences the number of females of child-bearing age present in the population now and this set of people in turn influences the number of babies born. In general, rates of births, migration and deaths only change marginally from one year to the next. 'Demographic momentum' defined by the age-sex structure of the population is a more powerful influence on the population at the next time point. Demographic momentum is usually associated with a continued population growth even when birth rates have fallen due to the large size of previous cohorts. Here we might regard demographic momentum in somewhat broader terms; that the size of any cohort is defined by the size of another cohort at a different time point.

To a degree we have control over our own demographic behaviour; whether to have children; when to move house and to where; and to mitigate premature mortality by adopting a healthy lifestyle. Changes to the population structure in an area will only occur slowly without the influence of other factors, a variety of which can affect the occurrence of demographic events. These factors, which are to a large extent inter-twined, may be environmental, sociodemographic, economic, technological and policy related. In terms of policy, national and local governments may actively seek to affect population size and age-sex structure. Sometimes population structure may be inadvertently affected by a policy concerned with something else. We give two contrasting examples below.

China's 'one child policy' is the largest scale direct attempt to control population size. The policy aimed to check a quickly growing population size but has been the subject of wide debate due to the manner in which the policy was implemented (see Greenhalgh, 2008 for more details). Whilst the rate of growth has been curtailed (Figure 4.10), the population in China has still been increasing and is projected to do so until around 2030. However, the knock-on effects of trying to reduce population growth through the one-child policy have been the impacts on gender ratio, social mix and imbalances in population structure. Figure 4.10 shows neontic dependency ratios first rapidly increasing then

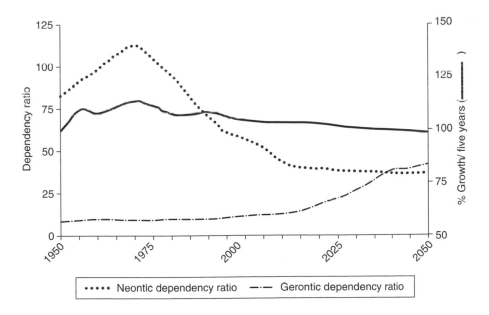

Figure 4.10   Dependency ratio and population growth, China, 1950–2050

*Source:* UN Population Division (2009b) World Population Prospects. *The 2008 Revision*

decreasing after the policy was introduced in the late 1970s. The bottoming out of the neontic ratios projected around 2010 will be because the smaller cohorts born during the 1970s and 1980s will be coming into working/child bearing age and are having fewer children themselves. Of concern is the rapidly ageing population as the large cohorts born during the 1950s and 1960s approach old age. After 2010, the gerontic dependency ratio increases at a rapid rate. This is likely to place great pressures both socially, with relatively small numbers of people looking after their more numerous elderly relatives, and economically, with a small working aged population challenged to create the wealth to support the infrastructures needed to support the elderly cohorts.

At a much more local level, Figure 4.11 illustrates the 2001 population in University ward in Leeds, West Yorkshire. (A ward is an area defined for the election of local councillors, and wards are widely used for collating statistics at a smaller scale.) Relative to the rest of the population in this area, there is a very large proportion of persons aged 15–19 (further age detail would show a predominance of persons aged 18 and 19) but the pyramid is dominated by the large spike of population aged 20–24; the peak age for students at a higher education establishment. There are relatively large populations up to ages in the mid-thirties most likely representing a postgraduate community. The location of a higher education establishment in the UK acts as the geographical destination for young adult migrants from within the country. In the UK, the majority of students do not study at their local institution but leave home to attend university. Students from overseas are also attracted to these locations. At a local level

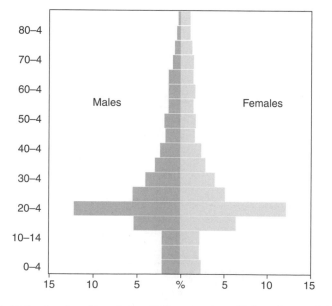

Figure 4.11   A distinctive local level population structure, University ward, Leeds, 2001
*Source:* 2001 UK Census

then, the age structure is affected by the presence of higher education establishments, the siting of student halls of residence and the availability of low-cost rental housing.

At both national and local levels, changes in population structure and needs for responses can be a side-effect of other policies so that changes in age structure occur as unintended consequences of policy initiatives. Using students as an example, we can suggest ways in which demographic rates and population structure and composition may be affected as a result of policies.

During the 1990s and 2000s student numbers were expanding within the UK. This was partly due to the then 'New Labour' government's goal to have 50 per cent of young adults in higher education (HE) and to increase participation from social groups and people in geographical areas who do not traditionally go to university. Since women with degrees and subsequent careers tend to have fewer children, the knock-on effect of increasing HE participation could be to suppress fertility rates in the future.

A change of government in the UK during 2010 led to the tightening up of visa regulations and a subsequent reduction in the number of international students able to attend the UK's HE institutions. Further government plans are likely to reduce the number of immigrants from countries outside the European Union. The ethnic mix of the UK's population in the short-term and ongoing will be affected.

Late in 2010, the government announced that from 2012 student tuition fees will rise substantially. This may affect both the numbers of students able to take up higher education places, the social mix (despite assurances that the less well-off will get assistance) and the location of the university which people choose. Perhaps more students will opt to study at their local university and live at home. The locations near to universities where students currently live may no longer have the large spike in young adult ages. The demand for rental housing and local goods and services may reduce.

Practitioners may respond to the age-sex structure by providing appropriate goods, facilities and other infrastructure. As with the events-structure inter-relationship discussed above, it is important to realise that the structure itself may be a response to decisions made. At national level, change in immigration policy affects the number of people coming into a country. At local level, the building of a new housing estate will lead to population growth, presuming the houses become occupied.

## Summary

- The population at any time point is a function of the previous population plus births which have occurred and minus deaths. The population size and structure will also be affected by subnational and international migration in and out of an area.

- Previous trends in population age-sex structure inform us how the current population has developed and how this relates to other social and health changes.

- Knowledge about how population size and structure will change in the future is important in a wide variety of applications including the likely demand for housing, schooling, employment, services and pensions as well as for business marketing.

- Population structures vary internationally between different countries and subnationally between different types of geographical areas and between different population subgroups.

- Within any area, population groups which may well have distinctive age-sex structures include: different ethnic groups; differently educated groups; and different social classes as defined by their occupations. These structures are formed by previous demographic behaviours and are predictive of future composition.

- Population ageing involves a redistribution over time of an area's population towards older ages. This usually includes a reduction in the proportion of young people in the population and a rise in the proportion of the population who are elderly. This represents 'relative ageing' but the numbers of elderly are relevant to the provision of services.

- Many developed countries have ageing populations and are moving towards superageing population structures as the baby boom generations approach retirement age. Developing countries though are ageing much faster and this will place pressure on populations, economies and care systems.

- Individuals have a degree of control over their demographic behaviour. At a population level, whether national or subnational, policy may affect the size and age-sex structure whether deliberately or as a side-effect of another decision.

## Recommended reading

For demographic definitions and calculation of age-sex structure see:

Rowland, D.T. (2003) *Demographic Methods and Concepts*. Oxford: Oxford University Press and Hinde, A. (1998) *Demographic Methods*. London: Arnold.

Rowland is also a useful source for methodologies for population projections.

On China's one-child policy see Susan Greenhalgh's 2008 book, *Just One Child: Science and Policy in Deng's China*. Berkeley: University of California Press.

On population ageing a useful review is:

Lutz, W., Sanderson, W. and Scherbov, S. (2008) 'The coming acceleration of global population ageing', *Nature*, 451: 716–719.

# 5

# ANALYSING FERTILITY AND MORTALITY

This chapter introduces the basic techniques of measuring and comparing fertility and mortality and how population specialists have developed their own language and techniques for studying fertility and mortality in a more rigorous or scientific, way. Questions and issues addressed in this chapter:

- How are fertility and mortality measured and why is it important to measure these accurately?

- What are the age patterns of fertility and mortality?

- What techniques can we use for comparing mortality levels of different places?

- Why is it important to distinguish between period versus cohort approaches to measuring demographic events?

Before we consider the basic methods of demographic inquiry, think about the following facts and figures (all data are taken from UNFPA *World Population Prospects: 2010 Revision*):

- For 2005–10, UNFPA estimated that the nation with the lowest Crude Death Rate (CDR) was United Arab Emirates with a CDR of 1 death per 1,000 population, the country with the highest CDR was the Central African Republic with a CDR of 18 per 1000.

- For the same time period, UNFPA estimated that Lesotho women had the lowest life expectancy at birth of 45.18, while Japanese women enjoyed the longest life span at 86.06.

- If we consider childhood mortality, children born in Singapore experienced the lowest under five mortality, with an estimated rate of two deaths under five per 1,000 live births between 2005–10, whilst children in Chad suffered the highest rate of 209 deaths per 1,000 live births. This means that one in five children born in Chad died before their fifth birthday.

- Women in the Hong Kong Special Administrative Region were estimated to have the lowest period of fertility from 2005–10 with a rate of 0.99, that is less than one child per woman. In contrast women in Niger had the highest fertility rate of 7.19 children per woman.

One of the issues that quickly emerges from any consideration of demographic events is the importance of accurate measurement. It is vital to be able to describe the trends and variations in population change, to distinguish between the contributions of the three components and to appreciate that clear description, let alone a convincing explanation, is fraught with difficulties. In this section, we review the main approaches to measuring fertility and mortality.

So far in this book we have mostly viewed global trends in population growth using the CBR and CDR. As outlined in Box 2.1, these rates are calculated with reference to the total population. For demographic events, the denominator refers to the population at risk of a particular event, in the case of the CBR and CDR, the 'at risk' population is the total population. However, it must be appreciated that both are poor measures of fertility and mortality since both are affected by the age structure of the population. Because CBR and CDR are crude rates which do not provide accurate measures of fertility and mortality it is important to treat them with care. Other measures provide more precise ways of measuring fertility and mortality, although they are all more demanding in terms of data requirements. More sophisticated demographic measures are age related, so that they relate births or deaths to the population 'at risk' of these events. For example, fertility measures relate the number of live births to the total female population aged 15 to 49, and sometimes might restrict this to married women in this age group. Similarly, mortality measures relate the number of deaths distinguished by age, to the population at risk of dying at those same ages.

# Measuring fertility

## Age specific fertility rates

The problem with CBR and CDR is that the denominators used do not provide a proper reflection of the precise population 'at risk' of giving birth or being more likely to die, and in this way are not in a strict sense measures of fertility or mortality. Fertility means the production of live-born infants which can only be accomplished by women during the period between menarche (the start of menstruation usually between ages 12–15) and the menopause (late forties to early fifties). By convention, the reproductive period is taken to be between the ages 15–49 in completed years (sometimes to 44). A better measure of fertility will therefore have the number of live-born infants as the numerator and the number of women aged 15–49 as the denominator. This is known as the general fertility rate (GFR, see Box 5.1). While superior to the CBR, it is more demanding in its data requirements, as the data used to derive the denominator must be distinguished by gender and age. However, fertility is not evenly distributed by age for women between ages 15 to 49 and GFR can still be improved upon to provide a more precise measure of fertility.

The next step to take in order to compute a more refined measure is to compute fertility rates for specific age groups of women. These are known as age specific fertility rates (ASFR) and are computed separately for each age group. Again the data need becomes more demanding as in order to compute ASFRs it is necessary to be able to divide up the number of live births according to the age of the mother. ASFRs are usually computed for five year age groups (15–19; 20–24; 25–29; 30–34; 35–39; 40–44; 45–49) hence the numerators are the number of births born to women in these age groups, and the denominators are the number of women in each age group.

## Total fertility, gross reproduction and net reproduction rates

ASFRs are an important first step in computing the most frequently used measure of fertility that is the total fertility rate (TFR). This is calculated by summing the ASFRs. TFRs give the total number of births a woman would have if she experienced the ASFRs throughout her reproductive life. It is important to remember that TFR is conventionally expressed per woman, while ASFRs are usually given for five-year age groups, hence after summing the AFRS, as they are for five year age groups it is necessary to multiply by 5. Details of how to compute GFR, ASFRS and TFR for the Philippines are given in Box 5.1

The TFR calculated for the Philippines for the year 2003 is 2.56, and we interpret this as follows: if a woman experienced the ASFR for the Philippines in 2003 for her entire reproductive life (from ages 15–49) the average number of births she will have is 2.56. It is not the average number of births women in the Philippines had in 2003 (or the average family size), and we will return to this important distinction when discussing period and cohort approaches to fertility. The TFR is often compared to replacement fertility, that this is the level of fertility required to replace a population. In low mortality societies, a TFR of 2.1 is assumed to equate to replacement fertility – i.e. each women needs to give birth to slightly more than two children. Two children would replace all mothers and fathers, but we need to allow for mortality as well, hence the .1 is to offset premature mortality. However, in high mortality societies 2.1 will not be sufficient, if a large number of babies born do not live to reproductive ages.

## Box 5.1   Computing general fertility, age specific fertility rates and total fertility rates

*General fertility rate*

$$GFR = \frac{\text{Live births in a year}}{\text{Mid-year population women aged 15–49}} * 1000$$

e.g. Philippines 2003

| | |
|---|---|
| Births: | 1664148 |
| Mid-year population women aged 15–49: | 21036292 |
| GFR: | 79 per 1000 women |

*Age specific fertility rates*

$$ASFR = \frac{\text{No of births in a year to women aged x to x+n}}{\text{Mid-year population women aged x to x+n}} * 1000$$

ASFRs can also be computed for single years of age:

$$\frac{\text{No of births in a year to women aged x}}{\text{Mid-year population women aged}} * 1000$$

e.g. Philippines 2003

| Age group | Births | Mid-year population women | ASFR per 1,000 women |
|---|---|---|---|
| 15–19 | 123036 | 4101200 | 30.0 |
| 20–24 | 472211 | 3765638 | 125.4 |
| 25–29 | 458341 | 3395119 | 135.0 |
| 30–34 | 340,211 | 3016055 | 112.8 |
| 35–39 | 193825 | 2626355 | 73.8 |
| 40–44 | 68400 | 2242623 | 30.5 |
| 45–49 | 8124 | 1889302 | 4.3 |

*Total fertility rate*

TFR is the sum of age specific fertility rates. Where the ASFRs are computed for single years of age this can be summarised as follows:

$$TFR = \frac{\text{Sum of ASFR}}{1000}$$

ASFRs are more usually computed for five year age groups, in this case the formula for computing the TFR is:

$$TFR = \frac{5 * \text{Sum of ASFR}}{1000}$$

e.g. Philippines 2000

NB this time the ASFR is calculated per woman, not per 1000 women

| Age group | ASFR |
|-----------|------|
| 15–19 | 0.03 |
| 20–24 | 0.13 |
| 25–29 | 0.14 |
| 30–34 | 0.11 |
| 35–39 | 0.07 |
| 40–44 | 0.03 |
| 45–49 | 0.004 |
| Total | 0.512 |

Sum of ASFR * 5 = 2.56

TFR – 2.56

Source: National Statistics Office, Republic of the Philippines
http://www.census.gov.ph/data/sectordata/2003/sr0620703.htm

The TFR can be refined further to give an indication of whether the female population is replacing itself. The gross reproduction rate (GRR) is similar to the TFR but is calculated for female births only. Formally this could be calculated by restricting the numerator in the ASFRs to female births, however, in practice a quicker method is to multiply the TFR by the proportion of the total number of births that are female. For example, in the Philippines in 2003, the sex ratio at birth was 105, in other words for every 205 babies born, 100 were girls. We can estimate the GRR thus

$$\text{GRR} = \text{TFR} * \text{proportion babies that are female}$$

$$= 2.56 * 100/205$$

$$= 1.25$$

The GRR measures replacement of a population in terms of the average number of daughters each mother will have. However, in order to get a more precise measure of replacement, we need to consider survival rates as well as fertility levels. A final fertility measure, the net reproduction rate (NRR) does just that as it adjusts the GRR with the survival rates of women from birth to the mean age of motherhood. A value of 1 for an NRR would indicate exact replacement, i.e. one daughter for each woman will survive into adulthood; below 1 indicates below-replacement fertility and above 1 is indicative of more daughters in future generations. The NRR will always be lower that the GRR; in low mortality societies the two indices will be very close, as mortality increases the differential between the NRR and the GRR will increase. However, NRRs (and TFR and GRR) should be treated with caution as the data from which they are computed usually refers to one year, and hence are subject to fluctuations in fertility and mortality.

## Measuring mortality

The CDR suffers from the same problem as the CBR, for while death may occur at any age, risks tend to be concentrated in early childhood and old age. This means that in a society with a very youthful age structure which is growing rapidly (over 1.5 per cent per year) CDR will appear low even if the risks of premature deaths are high. In an ageing society, where more people are aged over 60 than under 15, CDR will appear high even if the level of mortality is very low. This is illustrated in Table 5.1 which gives CDRs for eight different countries. While Afghanistan and Zimbabwe have the highest

Table 5.1    CDR for select countries, 2005–10

| Country | CDR per 1,000 population |
| --- | --- |
| Afghanistan | 17 |
| Zimbabwe | 15 |
| UK | 10 |
| Japan | 9 |
| USA | 8 |
| Thailand | 7 |
| Egypt | 5 |
| Mexico | 5 |

*Source:* United Nations Population Fund World Population Prospects, the 2010 Revision

CDR (and similar rates to that for Zimbabwe are found throughout Sub-Saharan Africa), Mexico and Egypt are at the bottom of the table (though as noted at the beginning of this chapter, UAE recorded a lower rate at 1 per 1,000), while both the UK and Japan have high CDRs. Hence we can see that CDR is a misleading way of comparing mortality in these countries.

## Age specific death rates

To measure mortality accurately, it is important to take age into account. The most straightforward way to do this is to compute age specific death rates (ASDR). These are derived in the same way as ASFRs:

$$\text{ASDR} = \frac{\text{No of deaths of individuals aged x to x+n}}{\text{Mid-year population aged x to x+n}} * 1000$$

However, the most commonly used ASDR is that for infants. In high mortality societies, the risk of dying during the first years of life is very high, and infant mortality is an important indicator of preventable mortality risks. For the first year of life, the infant mortality rate (IMR) is defined as follows:

$$\text{IMR} = \frac{\text{Deaths under one year of age}}{\text{Total live births in a calendar year}} * 1000$$

The IMR is therefore different from other ASDRs as the denominator is live births, rather than the mid-year population. Live births are used rather than mid-year population, as in the first year of life, deaths are not evenly distributed over the first 12 months, but are more likely to occur in the first days or weeks, hence the mid-year population would not provide a valid estimate of infants at risk of dying. Table 5.2 gives IMR for the same countries in Table 5.1.

Comparing IMR gives a very different picture of mortality levels in the seven countries, with stark contrasts between developing nations (Afghanistan, Zimbabwe, Egypt and Mexico) and the remaining four countries. However, it is important to bear in mind that in countries such as Afghanistan, accurate measurement of infant deaths and births will be considerably more problematic compared to countries at the bottom of the table.

As already noted above, infant deaths are not evenly distributed during the first year of life. In low infant mortality societies infant deaths will be very much concentrated during the first days and weeks of life and will almost exclusively occur as a result of premature birth or congenital malformations. However, in high infant mortality societies, the risks of

Table 5.2   IMR select countries, 2005–10

| Country | IMR per 1,000 live births |
|---|---|
| Afghanistan | 136 |
| Zimbabwe | 59 |
| Egypt | 26 |
| Mexico | 17 |
| Thailand | 12 |
| USA | 7 |
| UK | 5 |
| Japan | 3 |

*Source:* United Nations Population Fund World Population Prospects, the 2010 Revision

dying are not just associated with birth, but are persistent throughout the first year as infants are exposed to environmental risks. It is useful to distinguish between infant deaths that occur early (i.e. weeks one to four, referred to as neonatal deaths) and those that occur at older ages (occurring in weeks five to 52, or postneonatal deaths) as the causality of these deaths will be very different. Furthermore, risks associated with high IMRs are not just restricted to infants, as children under the age of five will be vulnerable. Hence one of the indices used by the UN in defining its Millennium Development goals is based on under-five mortality, not just infant mortality. Table 5.3 breaks down IMR and gives under-five mortality for three countries – Afghanistan, Zimbabwe and the UK – to illustrate how the different contribution of neonatal and postneonatal deaths contribute to the overall IMR. In the UK, when the rates are rounded to whole integers, infant mortality accounts for all under five mortality.

Table 5.3   Breakdown of IMR by neonatal and post-neonatal rates, 2000

| Country | IMR | Neonatal | | Postneonatal | | Under-five mortality |
|---|---|---|---|---|---|---|
| | | Rate | % of IMR | Rate | % of IMR | |
| Afghanistan | 165 | 60 | 36 | 105 | 64 | 257 |
| Zimbabwe | 73 | 33 | 45 | 40 | 55 | 117 |
| UK | 6 | 4 | 67 | 2 | 33 | 6 |

*Source:* United Nations Children Fund *State of the World's Children 2008* and *2002*

## Life tables and life expectancy

To compare mortality across the age range, the most common measure is life expectancy, denoted by $e_0$. Life expectancy is computed using life tables. Life tables are misleadingly named for they are really tables that express the chance of dying at particular ages and the numbers that would survive, if a particular set of ASDRs were held constant for a population. They were first constructed in the seventeenth century, developed in the eighteenth and perfected in the nineteenth. Their purpose is to act as a way of describing in some detail the pattern of mortality with age and for this reason their construction was pioneered by actuaries working for life assurance companies who wanted to estimate the level of premium required for a client to insure his or her life. Getting the calculation wrong (under-estimating the risk of premature death and thus charging the client too little) could have serious consequences for the financial viability of the companies concerned. Life tables are still used today for the purposes of computing insurance premiums and pension contributions. There is also a second use to life tables; they provide a simple model of a population with no migration and where the number of entry births and exit deaths are equal and where, in consequence, CBR = CDR.

Table 5.4 gives an example of a life table which can be used to illustrate its properties and the specific notation that is used for life tables. It is taken from the WHO Statistical Information System and gives a life table for women in Zimbabwe in 2000. It is an example of an abridged life-table, that is based on five year age groups, though if the data are available and a more precise measurement is needed (for example in the calculation of insurance premiums) life tables can be expanded for single years of age. In the abridged life table, the first year of life is always treated as a single year. Ages 1–4 are then grouped together in a four-year age groups, and all other age groups are for five years, except for the open-ended final age group. The first column gives the age group, column two gives the age ($x$) at the start of each age group and the third column gives the number of years ($n$) in each age group. The remaining three columns give three life-table functions (note that this is not a complete life table, further information on how to compute a life table is summarised in Box 5.2, on pages 81–3). The fourth column is headed $_nq_x$ and this notation signifies the probability of dying between ages x and x+n (by convention $n$ appears before the $q$ and $x$ after). For example, the probability of a woman dying between ages 25 and 30 is 0.11813 (i.e. over 10 per cent) and this is represented as $_5q_{25}$. The probability of dying in the first year of life is the same as the IMR. It is interesting to compare the number used by the WHO to compute this life table (0.05438 or 54 per 1,000) with that reported by UNICEF for the same year (73 per 1,000), while we would expect some difference as the latter is for both males and females and the former for females only, the differential illustrates the difficulty of accurately measuring mortality in less developed countries. The last two columns in Table 5.4 give, respectively, the number surviving to age x out of '100,000' births ($l_x$) and the life expectancy at age x in years ($e_x$). Taking the $l_x$ column first, this indicates that half of Zimbabwean women would be expected to die by their forty-fifth birthday ($l_{45}$ of 42304 is less than 50,000). The $e_x$ column shows that children surviving to their first birthday will have a greater life expectancy than at birth, as a result of the high risk of death in the first year of life for Zimbabwean babies. Yet by age 35, Zimbabwean women would be expected to live for another 26 years.

Table 5.4 Female life table, Zimbabwe 2000

| 1 | 2 | 3 | 4 | 5 | 6 |
|---|---|---|---|---|---|
| Age range | x | n | $_nq_x$ | $l_x$ | $e_x$ |
| <1 | 0 | 1 | 0.05438 | 100000 | 46.4 |
| 1–4 | 1 | 4 | 0.03141 | 94562 | 48.0 |
| 5–9 | 5 | 5 | 0.00714 | 91592 | 45.5 |
| 10–14 | 10 | 5 | 0.00505 | 90937 | 40.8 |
| 15–19 | 15 | 5 | 0.00733 | 90478 | 36.0 |
| 20–24 | 20 | 5 | 0.04322 | 89815 | 31.3 |
| 25–29 | 25 | 5 | 0.11813 | 85933 | 27.6 |
| 30–34 | 30 | 5 | 0.17797 | 75782 | 25.9 |
| 35–39 | 35 | 5 | 0.18527 | 62294 | 26.0 |
| 40–44 | 40 | 5 | 0.16647 | 50753 | 26.3 |
| 45–49 | 45 | 5 | 0.11118 | 42304 | 26.1 |
| 50–54 | 50 | 5 | 0.08488 | 37601 | 24.0 |
| 55–59 | 55 | 5 | 0.08883 | 34409 | 21.0 |
| 60–64 | 60 | 5 | 0.10380 | 31353 | 17.8 |
| 65–69 | 65 | 5 | 0.13522 | 28098 | 14.6 |
| 70–74 | 70 | 5 | 0.19702 | 24299 | 11.5 |
| 75–79 | 75 | 5 | 0.30114 | 19511 | 8.7 |
| 80–84 | 80 | 5 | 0.45474 | 13636 | 6.4 |
| 85–89 | 85 | 5 | 0.63591 | 7435 | 4.6 |
| 90–94 | 90 | 5 | 0.76381 | 2707 | 3.3 |
| 95–99 | 95 | 5 | 0.83885 | 639 | 2.4 |
| 100+ | 100 | | 1.00000 | 103 | 1.8 |

Source: WHO Statistical Information System, Life Tables for WHO Member States

Life tables and life expectancy are commonly misunderstood; $e_0$ does not equate to the average age of death recorded in a population or the maximum number of years that people may live. Rather, in the same way that TFR does for fertility, a life table takes a current set of death rates and applies it to a 'synthetic' cohort to calculate the average life expectancy of an individual if they experienced these death rates. In high mortality societies, $e_x$ is greatly affected by infant and childhood mortality. We can see that in Zimbabwe if a baby survives its first year then its life expectancy increases. However, the significance of infant mortality on life expectancy is often misunderstood, and it is often assumed, incorrectly, that a low life expectancy is indicative of the average age that people live to. It is a common misconception for example, that because life expectancy in many sub-Saharan African countries is less than 50, that no one lives beyond 50 in this part of the world. Moreover, some commentators have claimed that lack of longevity is directly associated with political ineptitude in this

## Box 5.2 Computing a life table

NB This box provides a summary of life table functions, notation and their computation. For a more detailed account of how to compute a life table see Rowland (2003)

| 1 Range | 2 x | 3 n | 4 $_nM_x$ | 5 nqx | 6 $l_x$ | 7 $_nd_x$ | 8 $_nL_x$ | 9 $T_x$ | 10 $e_x$ |
|---|---|---|---|---|---|---|---|---|---|
| <1 | 0 | 1 | 0.05653 | 0.05438 | 100,000 | 5,438 | 96,194 | 4635089 | 46.4 |
| 1–4 | 1 | 4 | 0.00800 | 0.03141 | 94562 | 2971 | 371120 | 4538896 | 48.0 |
| 5–9 | 5 | 5 | 0.00143 | 0.00714 | 91592 | 654 | 456323 | 4167776 | 45.5 |
| 10–14 | 10 | 5 | 0.00101 | 0.00505 | 90937 | 459 | 453539 | 3711453 | 40.8 |
| 15–19 | 15 | 5 | 0.00147 | 0.00733 | 90478 | 663 | 450732 | 3257914 | 36.0 |
| 20–24 | 20 | 5 | 0.00883 | 0.04322 | 89815 | 3882 | 439369 | 2807183 | 31.3 |
| 25–29 | 25 | 5 | 0.02511 | 0.11813 | 85933 | 10151 | 404286 | 2367814 | 27.6 |
| 30–34 | 30 | 5 | 0.03907 | 0.17797 | 75782 | 13487 | 345190 | 1963528 | 25.9 |
| 35–39 | 35 | 5 | 0.04084 | 0.18527 | 62294 | 11541 | 282618 | 1618338 | 26.0 |
| 40–44 | 40 | 5 | 0.03632 | 0.16647 | 50753 | 8449 | 232643 | 1335720 | 26.3 |
| 45–49 | 45 | 5 | 0.02354 | 0.11118 | 42304 | 4703 | 199763 | 1103077 | 26.1 |
| 50–54 | 50 | 5 | 0.01773 | 0.08488 | 37601 | 3192 | 180026 | 903314 | 24.0 |
| 55–59 | 55 | 5 | 0.01859 | 0.08883 | 34409 | 3057 | 164405 | 723288 | 21.0 |
| 60–64 | 60 | 5 | 0.02190 | 0.10380 | 31353 | 3254 | 148628 | 558883 | 17.8 |
| 65–69 | 65 | 5 | 0.02901 | 0.13522 | 28098 | 3800 | 130993 | 410255 | 14.6 |
| 70–74 | 70 | 5 | 0.04371 | 0.19702 | 24299 | 4787 | 109526 | 279262 | 11.5 |
| 75–79 | 75 | 5 | 0.07090 | 0.30114 | 19511 | 5876 | 82868 | 169736 | 8.7 |
| 80–84 | 80 | 5 | 0.11771 | 0.45474 | 13636 | 6201 | 52677 | 86868 | 6.4 |
| 85–89 | 85 | 5 | 0.18647 | 0.63591 | 7435 | 4728 | 25355 | 34192 | 4.6 |
| 90–94 | 90 | 5 | 0.28200 | 0.76381 | 2707 | 2068 | 7332 | 8836 | 3.3 |
| 95–99 | 95 | 5 | 0.40641 | 0.83885 | 639 | 536 | 1320 | 1504 | 2.4 |
| 100+ | 100 | | 0.55817 | 1.00000 | 103 | 103 | 185 | 185 | 1.8 |

Source: WHO Statistical Information System, Life Tables for WHO Member States

*(Continued)*

## Box 5.2   (Continued)

Columns 1 to 10 give:

  1  $x$      – age at start of age interval (years)

  2  $n$      – number of years in age interval (years)

  3  $_nM_x$   – ASDR for ages $x$ and $x + n$

  4  $_nq_x$   – probability of dying between exact ages $x$ and $x + n$

  5  $_np_x$   – probability of surviving between exact ages $x$ and $x + n$

  6  $l_x$    – number of persons alive at exact age $x$

  7  $_nd_x$   – number of persons dying between exact ages $x$ and $x + n$

  8  $_nL_x$   – number of person-years lived between exact ages $x$ and $x + n$

  9  $T_x$    – total number of person-years lived after exact age $x$

 10  $e_x$    – expectation of life from exact age $x$

$_nq_x$ is derivied from $_nM_x$ and the difference between the two relates to the dominator. Age specific death rates, $_nM_x$, are based on the mid-year population; however, the probability of dying during the year needs the population at the start of the year as its denominator. Hence the ASDRs or $_nM_x$ need to be corrected to give an estimate of the probability dying during the year as follows:

$$_nq_x = ( n * {}_nM_x ) / (1 + [ n * (0.5) * {}_nM_x ] )$$

$_np_x$ is the probability of surviving and is calculated by subtracting $_nq_x$ from one:

$$_np_x = 1 - {}_nq_x$$

NB for the last age group, the probability of dying is 1 and that of surviving is 0. $l_x$ is the number of persons alive at age x. The first value for $l_x$ is set at an arbitrary number (and is known as the radix) conventionally this is taken as a multiple of 1000 – usually 10,000 or 100,000. Subsequent values of $l_x$ are computed by multiplying the previous value of $l_x$ by $_np_x$:

$$l_x = l_{x-n} * {}_np_{x-n}$$

So, for example, the number of people alive age 20 is calculated by multiplying the number of people alive at age 15 by the probability of surviving between ages 15 and 20. $_nd_x$ is the number of deaths between ages $x$ and $x + n$ and is the difference between the number of people alive at two consecutive ages:

$$_nd_x = l_x - l_{x+n}$$

Hence the number of deaths between ages 15 and 20 is simply the difference between the number of people alive at ages 15 and 20. NB for the last age interval when $_nd_x = l_x$.

$_nL_x$ is the number of person years lived between ages $x$ and $x + n$. For a five year age group this is calculated as follows:

$$_nL_x = n/2 \, (l_x + l_{x+n})$$

where n=5 this can be written as $_nL_x = 2.5 \, (l_x + l_{x+n})$

For the age group 15–19, everyone who survives this age group lives 5 years, while those who die are assumed to live, on average 2.5 years, namely deaths are assumed to occur evenly over the five years.

There are two exemptions to this general formula for $_nL_x$.

1  For the first year of life the assumption that deaths occur evenly over the year is not reasonable, as infant mortality is concentrated in the early weeks. Hence we need to assume that those who die during the first year live on average slightly less than half a year, and the following estimation is often used:

$$_1L_0 = 0.3 \, l_0 + 0.7 \, l_{x+n}$$

2  For the last age interval an estimation of $_nL_x$ has to be used, a common one is $_nL_x = \,_nd_x / \,_nM_x$

$T_x$ is the total number of years lived from exact age $x$ and is computed by summing the $L_x$ column from the bottom upwards.

$$T_x = T_{x+n} + \,_nL_x$$

Hence the values of $L_{100}$ and $T_{100}$ are identical and $T_{95}$ is $T_{100}$ plus $L_{95}$ etc.

The final column is life expectancy, and is computed quite simply by dividing the total number of years lived from age $x$ ($T_x$) by the number of people alive at age $x$:

$$e_x = T_x / l_x$$

Hence $e_0$ is $T_0$ divided by $l_0$ or the radix: 4635089/100,000 = 46.4

part of the world. But this cannot be verified by lifetable data; life expectancy is not a measure of maximum lifespan, but an indicator of average mortality risk (Gorman, 2011). If we return to the Zimbabwe life table, as discussed above, on average women at age 30 could expect to live another 26 years (to age 56, longer than life expectancy at birth). Thus even though many Zimbabweans die before their 50 birthday, many do live beyond this age, at the time of writing, the country was still ruled by the octogenarian Robert Mugabe.

## Age patterns of fertility and mortality

It will be obvious by now that both fertility and mortality are strongly influenced by age, that there are clear age-specific patterns for both. In this section, we explore these age variations in more detail.

Let us begin with fertility. Figure 5.1 illustrates the typical form of the age-specific pattern with fertility at its highest among women in their twenties (20–24 or 25–29) with lower rates among teenagers and particularly low rates among women in their forties.

Figure 5.1 also reveals the differences between countries. Zimbabwe and Egypt have the highest TFRs (above 3) and this is associated with very high fertility for women during their twenties, though fertility peaks earlier in Zimbabwe than Egypt. Mexico and

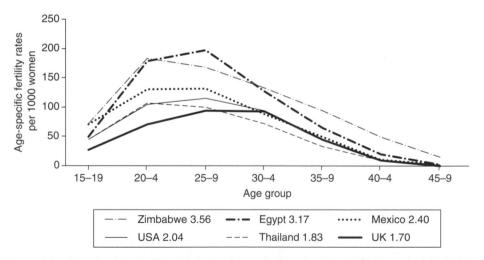

Figure 5.1   Age specific fertility rates for six example countries 2000–5

*Note:* TFR is given in the key

*Source:* UN Statistics Division (2008) Gender Info 2007 Retrieved June 2010 from http://unstats.un.org/unsd/demographic/products/genderinfo/

the US have similar ASFR at older ages, but at younger ages Mexico's fertility is higher. Thailand and the UK illustrate the difference between early and late patterns of fertility, while both have similar TFRs (compared to the other countries), Thailand has a younger fertility profile – peaking in ages 20–24, compared to the UK where women aged 30–34 have the highest fertility.

The same countries (with the exception of Thailand) are shown in Figure 5.2, which gives age specific death rates. Note that the vertical axis has been logged in order to illustrate the pattern more clearly and life expectancy is given in the key. We can see that four of the five countries have the classic tick-shaped profile, but that of Zimbabwe is quite different with a dramatic increase in mortality from ages 20–24 onwards, which does not level out until ages 40–44. This bulge in mortality is a result of an AIDS epidemic in Zimbabwe causing premature death among Zimbabwean adults. This is discussed in more detail in Chapter 9.

It is important at this point to emphasise the significance of gender differentials in mortality profiles. Figure 5.3 shows gender and age specific death rates for two of the example countries, Zimbabwe and UK. Male excessive mortality is more common than female excess, though the rates for Zimbabwe illustrate higher female mortality between the ages of 15 and 30 which can be attributed to maternal mortality, though male mortality peaks at a higher level at ages 35–39. There is little difference in mortality by gender at childhood, though there are some societies where this is observed. In the UK, there is a noticeable increase in male mortality after age 15 which is due to higher male mortality from road traffic accidents and other high-risk activities. At older

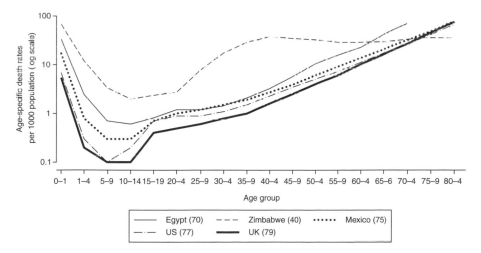

Figure 5.2   Age specific death rates for five example countries 1999–2004

*Note:* Life expectancy (2000–05) is given in the key

*Source:* UN Statistics Division Demographic Yearbook 2005 and 2004

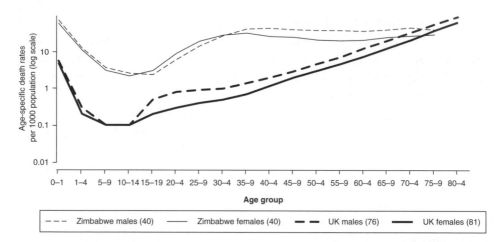

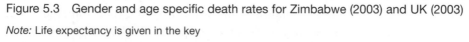

Figure 5.3    Gender and age specific death rates for Zimbabwe (2003) and UK (2003)

*Note:* Life expectancy is given in the key

*Source:* UN Statistics Division (2006) Demographic Yearbook 2005

ages in both countries, men have higher death rates, though the differential remains constant with age.

The five example countries illustrate some of the contemporary global differences in mortality and the range of experiences between different societies today in terms of their mortality levels and age profile of mortality risks. We shall conclude this section by looking at the extreme mortality situations as estimated by the World Health Organisation for the year 2009. The WHO annually produces a series of key health, morbidity and mortality indicators, one of which is life expectancy; it also computes life tables for each member country. In 2009, the highest life expectancy was recorded for Japanese women who had an $e_0$ of 86.5 years. At the other end of the range Lesotho recorded one of the lowest life expectancies, with 49.7 years for women and 46.3 for men (note that these figures are higher than the UNFPA estimates quoted at the beginning of the chapter). Figure 5.4 illustrates the mortality profiles of these two populations using $_nq_x$. It captures the variation in mortality experiences in the world at the beginning of the twenty-first century. For comparison, the $_nq_x$ curve for England and Wales 1838–54 is also shown with its female $e_0$ of 42 years. The difference between Lesotho and Japan is striking and illustrates the huge differences in life chances for babies born in these two countries in the early years of the millennium. There are also interesting comparisons between Lesotho and England. The former, like Zimbabwe above, displays the distinctive sharp rise in mortality in the adult years, associated with HIV-AIDS. In contrast in mid-nineteenth-century England, infant and child mortality made a bigger contribution to low life expectancy.

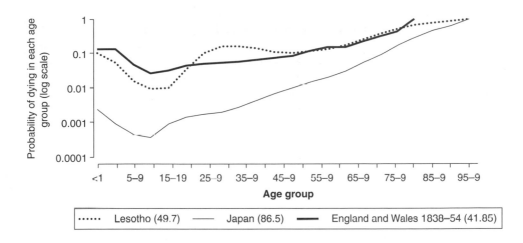

Figure 5.4   Female $_nq_x$ curve for Japan and Lesotho, 2009 and England and Wales 1838–54

Life expectancy is given in the key

*Source:* WHO (2011) Life tables for member states. Retrieved from www.who.int/healthinfo/statistics/mortality_life_tables/en/

Data for England and Wales were provided by Professor R.I. Woods

# Measuring fertility change: proximate determinants of fertility

It is apparent from the data on fertility presented in this chapter that the level of fertility recorded in populations at different times and geographies, is considerably lower than the biological limit of human reproductivity. The 'highest' fertility level consistently recorded for a population was observed for the Hutterites, an anti-Baptist religious group who live in North America. In the first half of the twentieth century, the TFR for the Hutterites averaged 8.9. Yet even this level of fertility is below that which could be theoretically achieved. In order to understand variation in fertility in different societies, it is useful to identify what we might term the proximate, or intermediate, determinants of fertility (Bongaarts, 1978). These proximate determinants focus attention on both the biological and cultural controls of conception and childbirth, see Box 5.3. For example, the proportion of the population that is married, and sexual behaviour within marriage, are cultural practices that effectively limit fertility. In historic European societies, older ages of first marriage and high rates of spinsterhood/bachelorhood limited fertility, while in societies where people marry young universal marriage practices do not restrict fertility in the same way. Lactational amenorrhea refers to the time when women are breastfeeding and do not ovulate or have menstrual periods. While this is clearly a biological function, cultural practices that influence the

---

### Box 5.3   Proximate determinants of fertility

| Indirect determinants | ⟶ | Proximate determinants | ⟶ | Fertility |

Indirect determinants ⟶ Proximate determinants ⟶ Fertility
Socioeconomic                 Proportion of women married
Cultural and                  Frequency of intercourse
Environmental factors         Post-partum abstinence
                              Lactational amenorrhoea
                              Contraception
                              Spontaneous abortion
                              Induced abortion
                              Natural sterility
                              Pathological sterility

---

length and frequency of breastfeeding will either extend or curtail this determinant. Particularly when combined with cultural practices of post-partum abstinence, lactational amenorrhoea will extend birth intervals and so limit women's achieved fertility. These determinates are more common in many developing societies particularly in Sub-Saharan Africa, thus 'development' initiatives that might restrict breastfeeding will result in increases in fertility. Frequency of intercourse is also partly determined by age (younger couples have more active sex lives), but also cultural practices. For example, in nomadic societies, spousal separation will reduce couples' sexual activity. The number of sexual partners is also important; in polygamous societies, men have more sexual partners, but women have less frequent sexual intercourse thus suppressing their fertility. Moreover women in polygamous relationships may be more at risk of pathological sterility, as it is caused primarily by the transmission of sexually transmitted disease (see Gould, 2009: 136–53 for further discussion of the relationship between fertility and culture).

The importance of the proximate determinants conceptualisation of fertility is that it identifies different mechanisms by which fertility can be controlled and why fertility is lower than might be expected without access to modern contraception. Moreover these determinants can be approximated by a series of indicators, which are outlined in Box 5.4. Further details on how these indices are computed can be found in Rowland (2003: 226–9).

## Comparing mortality levels

Although life table measures can provide an excellent way of looking at levels of mortality in different countries or specific populations, they can be rather demanding in terms

## Box 5.4  Indices for proximate determinants

The index of marriage: $C_m$. This refers to the proportion of women who are married, and equals 1 if all women of reproductive age are married and 0 if none are married.

The index of contraception: $C_c$. The index considers the prevalence and effectiveness of contraception and equals 1 if there is either no use of contraception or if its use is completely inefficient, and 0 if all women of reproductive age use 100 per cent efficient methods.

The index of postpartum infecundability $C_i$. The index equals 1 if there is no breast-feeding and no postpartum abstinence of sexual relations and 0 if the duration of infecundability is infinite.

The index of induced abortion: $C_a$. This ranges from 1 if there is no induced abortion to 0 if all foetuses are aborted.

Each of these indices reduce the total fertility rate. If we take TF as a measure of total fertility, that is fertility that would be achieved in the absence of any of the proximate determinants, then the relationship between achieved fertility, i.e. the TFR, and total fertility can be written as follows:

$$TFR = TF * C_m * C_c * C_i * C_a$$

In other words, the closer the indices are to 1, then the smaller the differential between TFR and TF.

---

of the data required to compute a life table. What we need is a simpler single number measure that can be used to compare both large and small populations but which also removes the potentially distorting effects of differences in age structure. The standardised mortality ratio (SMR) fits these requirements quite nicely. It is one of a group of measures which use indirect standardisation; that is the method applies a selected mortality schedule as a standard pattern of mortality to the age structures of the population(s) to be compared. An example of how to compute a SMR is shown in Box 5.5. This example compares the death rates of England and Wales, Northern Ireland and Scotland. From the crude death rates, it would appear that Northern Ireland has the lowest mortality. However, Northern Ireland also has a younger age population than England and Wales and Scotland, and so we need to consider how this affects the CDR. We take the age-specific death rates for England and Wales as a standard set of death rates and apply these to the population age structures for Scotland and Northern Ireland. This generates columns four and five, that is the expected number of deaths in Scotland and Northern Ireland if both countries experienced the ASDRs for England and Wales in 2001. The SMR is calculated by dividing the number of observed deaths by the number of expected deaths.

## Box 5.5    Calculation of standardised mortality ratio

CDR and total number of deaths, England and Wales, Scotland and Northern Ireland, 2001

| Country | CDR 2001 | Total number of deaths 2001 |
|---|---|---|
| England and Wales | 10.1 | 530300 |
| Scotland | 11.3 | 54400 |
| Northern Ireland | 8.6 | 14500 |

*Source:* ONS Population Trends (2008)

Calculations for SMR, Northern Ireland and Scotland

| Age group | Population 2001 | | Age-specific death rates England and Wales 2001 | Expected deaths | |
|---|---|---|---|---|---|
| | Scotland | Northern Ireland | | Scotland Col 1* Col 3 | Northern Ireland Col 2* Col 3 |
| | Column 1 | Column 2 | Column 3 | Column 4 | Column5 |
| 0–4 | 276874 | 115238 | 1.047 | 290 | 121 |
| 5–24 | 1261668 | 494300 | 0.294 | 371 | 145 |
| 24–44 | 1480261 | 489195 | 0.993 | 1470 | 486 |
| 45–64 | 1238308 | 363209 | 5.430 | 6724 | 1972 |
| 65–84 | 716545 | 200023 | 37.505 | 26874 | 7502 |
| 85+ | 88355 | 23301 | 164.559 | 14540 | 3834 |
| Total | 5062011 | 1685266 | | 50268 | 14060 |

*Source:* ONS Population Trends (2008)

If we sum the expected deaths we can compare this figure with the actual number of deaths observed in the two countries in that year. This is exactly what the SMR does; it divides the number of observed deaths by the number of expected deaths:

$$SMR = \frac{\text{Observed deaths}}{\text{Expected deaths}} *100$$

Conventionally the SMR is multiplied by 100. The SMRs of Scotland and Northern Ireland are:

$$SMR\ Scotland = \frac{54400}{50268} *100 = 108$$

$$SMR\ Northern\ Ireland = \frac{14500}{14060} * 100 = 103$$

In both countries, the SMR is greater than 100. This means that the level of mortality in Scotland and Northern Ireland is greater than that in England and Wales. Hence while the CDR for Northern Ireland is lower than that for England and Wales, standardising for age illustrates that the overall level of mortality is in fact higher. For more detail on how to compute SMRs see Rowland 2003.

SMRs are often used to compare the mortality experience of subgroups of a population that may have very different age structures, for example to compare different occupational groups or geographical areas. They are also used to look at changes over time and in this case one particular year is selected to be the standard against which other years can be compared. We shall see examples of SMRs being used in Chapter 9 which considers health inequalities.

## Period versus cohort approaches

We have already discussed the term 'period' as applied to fertility rates when discussing TFR. However, it is important in looking at the structure of populations, especially their component fertility, mortality and migration patterns, to draw a distinction between period and cohort approaches. The final section of this chapter takes a brief look at these differences as they relate to mortality and fertility.

A period approach considers what is happening at a given point in time. As outlined in Chapter 3, the best example is a population census which is a record of the individuals at one specific point in time, for example the UK census day in 2011 was 27 March. The characteristics of the entire population are recorded: numbers, age, sex, place of birth, occupation, marital status, education, and so on are commonly included. Since it is normal to use census data as the basis for the denominator, or the 'at risk' population, for mortality and fertility rates, it should not be surprising that CBR, CDR, TFR, GFR and most life tables are period measures. They relate to fertility or mortality experiences that occurred during a specific short time period, usually one, three or five years depending on how the numerator was constructed. For example, to compute a life table, it is possible to take the average of the number of deaths over a three-year period for the numerator and a population census for one specific year for the denominator. Yet there is a problem with period measures computed in this way as they ignore the effects of change. For example, Table 5.4 suggests that a baby girl born in Zimbabwe in 2000 might expect to live, on average, a further 46.4 years, but this is based on the assumption that she will experience the same level of mortality when aged 40, for example, as a 40 year old experienced in 2000. Clearly this is an unreasonable assumption if mortality declines (or increases) significantly. It could be that adult mortality declines in Zimbabwe over the next 40 years, and that the real life expectancy of the 2000 birth cohort will be greater than 46 years, but we cannot know this.

An alternative approach to studying fertility and mortality is to take a cohort approach. Cohort studies take a group of people with some shared experience and trace their subsequent history. For example we can define birth cohorts, school-leaving cohorts, marriage or divorce cohorts, in-migration cohorts, individuals first diagnosed as having a specific disease at a particular time (cancer studies often take this approach) and so forth. Once the group has been defined, then its subsequent experiences are tracked, but this is not always a straightforward matter. It is partly because of their complicated structure that cohort studies are only used in special circumstances; however, it must also be emphasised that research based on such material has proved particularly revealing in work on life histories, migration histories and in epidemiology where morbidity and mortality can be considered jointly.

One of the best ways of illustrating the differences between period and cohort approaches and of demonstrating the respective strengths and weaknesses is via the lexis diagram. Figure 5.5 shows an example of a lexis diagram which is named after its originator, Wilhelm Lexis. The horizontal axis on the lexis diagram depicts calendar time: the specific points refer to dates (e.g. 1 January) while the intervals are calendar years. The vertical axis represents time in the study, usually age but it could be time since leaving school, arriving in a country or being diagnosed with an illness. Points on the vertical axis are anniversaries of these events. Diagonal lines on the grid represent

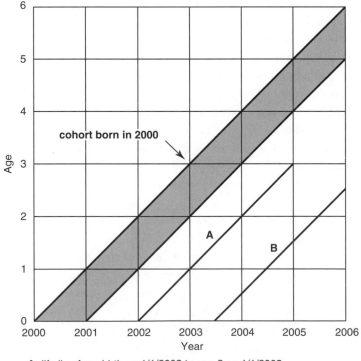

**A:** life line from birth on 1/1/2002 to age 3 on 1/1/2003
**B:** life line from birth on 15/6/2003 to age 3½ on 1/1/2006

Figure 5.5   Lexis Diagram

individual lives. For example for a birth cohort, a baby born on January 1 2000 will be one on 1 January 2001, etc. Each individual diagonal line is called a 'life line' or a 'line of life'. In Figure 5.5, the life lines of all individuals born in the year 2000 are shaded to represent the birth cohort for 2000. Period measures involve using the vertical cross-sections on the diagram, while horizontal cross-sections will give information on age groups. However, cohort analysis uses diagonal cross-sections, for example, for one birth cohort such as that illustrated for the year 2000 in Figure 5.5.

The main limitations in using cohort approaches are due to data restrictions. For example, completed fertility of birth (or marriage) cohorts will indicate women's achieved fertility and is a more robust way of examining fertility change over time. However, in order to compute completed fertility, it is necessary to wait until women have reached their late forties or early fifties, hence it is not possible to gauge what is happening to contemporary completed family size. Figure 5.6 gives age-specific fertility rates computed for different birth cohorts of women in England born between 1950 and 1980. In other words, the graph for 1950 gives the ASFR for women born in 1950 when they were under 20, 20–24, etc. Not all the cohorts are completed, as women born in 1980 are only observed until 2005 when they are aged less than 25.

We can see from Figure 5.6 that women born in the later years are having their children at older ages. For example fertility for the 1950s cohort peaked at ages 20–24 and declined thereafter, while for women born from 1955 onwards, fertility peaked at ages 25–29. For women born in more recent years fertility has been higher at older ages. However, the completed fertility for women has remained quite stable over the thirty-year

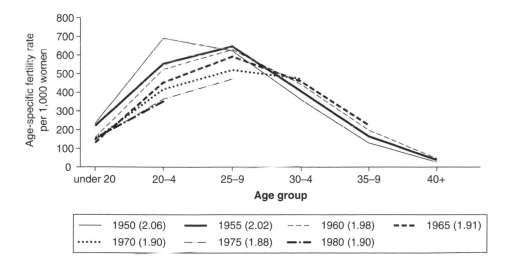

Figure 5.6   Age specific fertility rates by birth cohort, England. NB completed family size is given in the key (this is estimated for years 1965 onwards)

*Source:* © Government Actuary Department (2006) Fertility Assumptions Retrieved October 2011 from www.gad.gov.uk/Demography%20Data/Population/2006/methodology/fertass.html

period, falling from 2.06 in 1950 to a projected 1.90 for women born in 1980. Hence the variation in timing of births has not resulted in significant shifts in completed fertility, and overall measures of completed fertility derived from cohort data are less variable than those derived from period data.

## Summary

- The most basic demographic measures of fertility and mortality – crude birth and death rates – are not suitable for comparing fertility and mortality between different geographical areas or time periods, as they do not take into account the age structure of the population and are not defined by the age of those at 'risk' of giving birth or dying.

- The general fertility rates, total fertility rate, gross reproduction rate and net reproduction rate can be used to measure and compare fertility. The Total Fertility Rate is the most commonly used measure and indicates the number of children a woman will have if she experiences current age-specific fertility rates during her reproductive years, it is not a measure of average family size.

- Life tables are used to compute life expectancy, the most commonly used measure of mortality. Life expectancy at birth indicates the number of years an individual can expect to live if she experiences current age-specific mortality, it is not a measure of average age at death. Life expectancy at birth will be lower than life expectancy at age 1 or 5 if infant and childhood mortality are high.

- A more straightforward way of comparing mortality (and which can also be applied to other demographic events) is to standardise rates. For example, the standardised mortality rate uses a standard set of age-specific death rates and applies these to different population age structures for the geographical regions to be compared.

- As an alternative to period measures, such as the Total Fertility Rate, a cohort approach to data collection and analysis may be used. Cohort analysis of fertility, for example, will provide a more stable analysis of fertility trends over time. However, the delay in collecting this data (we have to wait for women to complete their fertility) means that we have to rely on old data, or estimate rates for more recent cohorts, hence period measures are favoured as they are more up to date.

## Recommended reading

This chapter has given a brief introduction to demographic techniques which will enable students to understand their use throughout this book. It is not intended as a definitive guide on how to compute demographic methods. For these techniques, there are a number of detailed texts available. For example Donald T. Rowland (2003) *Demographic Methods and Concepts*, Oxford, Oxford University Press gives detailed examples of all the techniques mentioned in this chapter. Colin Newell (1988) *Methods and Models in Demography,* London, Belhaven, gives a straightforward introduction to basic demographic methods. For a more advanced approach see Andrew Hinde (1998) *Demographic Methods,* Oxford, Oxford University Press.

This chapter has used a variety of demographic data sources that are available on the web, these include:

UN data website: http://data.un.org/Default.aspx. This website brings together the main UN data sources and provides an easy to search facility for key demographic indicators.

The UN Population Fund *World Population: The 2010 Revision* can be found at: http://esa.un.org/unpd/wpp/Excel-Data/mortality.htm

WHO statistics (in addition to those on the UN website) can be found at www.who.int/research/en/

The UN *Demographic Yearbooks* can be found on the UN Statistics Division webpages: http://unstats.un.org/unsd/demographic/products/dyb/dyb2.htm

The State of the World's Children Annual Reports are published by UNICEF: www.unicef.org/publications/index.html

For data on UK population see the ONS statistics hub at http://www.statistics.gov.uk/hub/index.html and in particular the publication Population Trends which can be downloaded at: www.ons.gov.uk/ons/rel/population-trends-rd/population-trends/index.html

The US Census Bureau provides comprehensive coverage, including an international database: www.census.gov/

# 6

# MIGRATION

Migration – the movement of people to live in a different place – is the third component of population change which, together with fertility and mortality, changes the characteristics or composition of an area's population. In particular, migration changes the distribution of the population; it changes the geography of the population. For this reason, migration is one of the central interests of population geographers, as well as demographers, sociologists, economists and other social scientists.

This chapter reviews definitions of migration, models of migration processes and theoretical approaches that aim to understand the causes and consequences of migration. The chapter also considers the ways in which researchers measure and monitor migration.

The chapter addresses the following questions:

- Why is migration important?
- What is migration?
- What makes migration happen?
- How is migration measured?
- How does migration vary between places and times?
- What approaches do population geographers take to studying migration?

## The importance of migration

Migration, which we will for now understand broadly as the movement of people to live in a different place (how to define migration more specifically is discussed in the next section), is important for people, places and societies. In terms of the demographic components of population change, migration is the third component which, together with births and deaths, alters the population – size, structure and composition – of a place (see Chapter 4 on population structures). This changes the nature of places and societies. Government, services and business have to respond and adapt to population change due to migration in order to meet the needs of the population, represent them politically, and, in the case of business, ensure profitability.

Migration is also important because it can be a significant event (and process) in individuals' lives. To move house, be it over a short distance for reasons of family or to another country for work, entails a lot of planning, decision making and action; it has potential risks and rewards; and involves social, emotional and financial upheaval. It is likely that you have moved house, even moved region or country, and it is likely that this event has affected your life and the lives of your family and friends.

Migration can be interpreted as a reflection of social, economic, political and environmental events and processes at global, national and local scales. It is therefore a valuable lens through which to understand social change. For example, Egyptian refugees attempting to cross the border from Libya into Tunisia in March 2011 were reacting to political upheaval in North Africa; in 2005 thousands of people were forced from their homes in New Orleans in the US following devastation brought by Hurricane Katrina; every year agricultural workers migrate seasonally around Australia following the harvest trail; and over the last three decades people in Britain (and other more developed countries) have been moving away from cities for lifestyles offered by more rural areas. Migration represents the reaction of populations to social change and, simultaneously, can bring about social change.

The selection of statistics below illustrates the numerical significance of migration (International Organisation for Migration, no date). By the end of this chapter you will think of these figures as more than numbers; as a representation of complex processes of decisions and experiences of migrants locally, nationally and globally.

- In 2008, there were 214 million international migrants worldwide (UN, 2009).

- In 2008, international migrants represented 3.1 per cent of the world's population (UN, 2009a).

- In 2009, there were 15.2 million refugees in the world (UN High Commissioner for Refugees, 2010).

- In 2009, international migrants sent US$414 billion in remittances (money sent from country of immigration to recipients elsewhere) (World Bank, 2010a).

- In an average year, one in every ten people will move house within Britain (UK Census, 2001).

## Defining migration

Defining migration would seem very straightforward! Migration is an idea that is used in everyday conversation, in the media and is something that affects most of us at some point in our lives. So, the central meaning of migration is clear: it is the movement of a person to live in a different place.

However, if we begin to think about the specifics of the definition of migration, it becomes more challenging. What is meant by 'to live in'? What constitutes a 'different place'? These distinctions are important because they indicate different types of migration, with different causes, consequences and experiences, and they also affect what methods are used to investigate migration.

Migration can be differentiated along a number of lines. Clarke (1972: 130) commented that:

> The extreme diversity of migration in cause, duration, distance, direction, volume, velocity, selectivity and organization prohibits simple classification. We read of seasonal, temporary, periodic, and permanent migrations, of spontaneous, forced, impelled, free and planned migration, as well as of internal, external, inter-regional, international, continental and inter-continental migrations. Obviously, no typology satisfactorily incorporates all types of human migration.

So, migration can be sub-divided along numerous lines of classification. Here we will review four of these which are interlinked and cover some of the fundamental aspects of migration: the crossing of geographical boundaries, the timing of the move, the direction of the move and the reason for the move.

## Crossing of geographical boundaries

Categorisation of migration by the crossing of geographical boundaries most obviously divides moves into international migration and internal migration. International migration is a residential move that crosses a national boundary. So, if you were to go to live in Australia having previously lived in Britain, you would be an international migrant. From the destination country's perspective (in this case Australia), you would be an immigrant; from the sending country's perspective (in this case Britain), you would be an emigrant.

International migration has become more common as international transport and communications have developed, economic and trade links have globalised, families have become transnational and laws have adapted to enable movement of people across international boundaries. In some parts of the world, such as within the European Union, there is freedom of movement for residence and work, according to European Parliament and Council Directive 2004/38/EC of 29th April 2004 on the right of citizens of the Union and their family members to move and reside freely within the territory of the member states. However, most international migration is subject to regulations that mean migrants have to meet certain criteria before being allowed to live and work in the destination country.

Indeed, international migration policies are the subject of much debate and positions range from those advocating a closure of borders (e.g. MigrationWatchUK) to those advocating open borders (e.g. Riley, 2008). These are political debates about the rights that individuals have to live in countries outside that which they were born in. Rights to live in a country, and also to work, to access housing, to use services such as education and healthcare, to participate politically (to vote), are generally linked to citizenship (or nationality). Commonly, citizenship is gained through birth in a country (*jus soli* or birthright citizenship), parental (or grandparental) citizenship (*jus sanguinis*), spouse citizenship or application for citizenship after a period of residence in the country (naturalisation). A person can be a citizen of one country, more than one country or no country (stateless persons).

As international migration has become more common over the last century, so too have multiple citizenships and people's connections with more than one place (country). The field of study concerned with cross-national interconnections is transnational studies (Glick Schiller and Faist, 2010). Concepts of transnationalism question the meaning and role of the nation state (Cohen, 2006; Samers, 2010).

The question of belonging, in a legal sense, to the nation in which you reside, is particularly pertinent for irregular or undocumented international migrants. Precisely because they are undocumented, relatively little is known about the numbers, characteristics and experiences of undocumented migrants. Some countries, including the US, Italy and Spain, have held immigration amnesties, where undocumented migrants are given legal status and rights to reside and work. Perhaps the most comprehensive study of undocumented migration is the Mexican Migration Project which surveys documented and undocumented migrants from Mexico to the US and vice versa in order to understand their characteristics, behaviour and experiences (Durand and Massey, 2004).

Though boundary crossing is most evident for international migration, residential moves within countries also involve boundary crossing. Subnational migration is known as internal migration and is also referred to as residential mobility or spatial mobility (not to be confused with social mobility, which refers to changes in economic/ professional/class status). So, if you moved within England from Manchester to London you would be an internal migrant. From the destination city's perspective (in this case London) you would be an in-migrant; from the sending city's perspective (in this case Manchester) you would be an out-migrant.

Internal migration can result in the crossing of a boundary which may have implications for the migrant. This boundary may be administrative, related for example to housing registration and local tax payments; electoral, based on boundaries of political representation, such as states in the US; or cultural, for example a move within Belgium from Flanders to Wallonia would mean a change in official language from Dutch to French.

Internal migration is widely experienced and in most societies reflects the desires and abilities of people to live in different parts of a country. Ability to move house within a country is not usually restricted by laws or regulations (though it is subject to other barriers and constraints). In some places, however, official internal migration is legally restricted. This is the case in China where residential location is controlled via the Hukou system. The example of internal migration in China is expanded in Box 6.1.

## Box 6.1    Internal migration in China: restrictions of the Hukou system

China is a unique context for migration due to its geographical and population size, its politics and its economic situation.

China, as a state, can be considered a closed migration system because there are very few international migrants. However, migration within China – specifically from rural to urban areas – can be seen as comparable to international migration elsewhere in terms of the drivers of the migration and the rights and experiences of migrants.

China has a long history of distinction between urban populations and rural 'peasant' populations. This distinction was re-enforced in the Communist period of the People's Republic of China since the 1950s by laws to prevent residential mobility. This has operated via the population and housing regulation system, the Hukou system.

However, the rural population's desire for a better standard of living coupled with the need for labour in urban areas has meant there have been large flows of migrants from rural to urban areas. This has particularly been the case since economic reforms began in 1979. The migrants, know as a 'floating population' total around 80 million people.

Rural migrants to urban areas have commonalities with 'illegal' or 'undocumented' migrants in other countries. Whilst official urban residents have rights to state support in the realms of work, housing, transport, education, health and more, rural migrants do not share these rights.

The legal basis for this division was first laid by a June 1955 State Council directive on establishing a system of household registration rules. Those not registered in the area they are residing – the case for most rural migrants to cities – do not have the same entitlements as registered residents.

Some amendments to policy have eased the situation of migrants: since the late 1980s temporary household registration has been available and application for urban registration in the form of the 'blue hukou' has been possible since the early 1990s.

Nevertheless, the conditions for migrants are poor. Although independent (non-state) workers unions exist, migrant workers fear joining them because of the risk of being identified and losing employment.

In sum, it can be argued that 'rapid growth has been openly acknowledged to be the result of the cheaply-recompensed drudgery of outsiders ... The purpose of the state was to make the peasantry a potential underclass, ready to be exploited to fulfil the new state's project of industrialization' (Solinger, 1999: 458, 465).

However, there are several indications that the situation may become less divisive and unequal: reforms are affording more rights to migrant workers; improvement of workers' rights more generally will have side-effects for migrant workers; economic engagement between China and Western states will encourage policies that will benefit migrant workers.

Source: Solinger (1999)

Boundary crossing in internal migration studies has traditionally been referred to as 'distance of move'. Distance is a useful migration measure but can be a difficult concept for comparisons between countries because of their differing sizes and, thus, the differing meaning of distance. For example, a move from the north to the south of the state of California, in the US, would be a greater distance than moving from the very south of England to the very north of Scotland or three times the distance of a move from London to Paris. Thus, it is important to interpret distance of migration in relation to boundary crossing.

This concept of distance of migration has traditionally been important for population geographers because it has been theorised to be linked to the reason for moving. Specifically, long distance moves have been traditionally seen as economically motivated, such as for a new job. In contrast, short distance moves have been seen as lifestyle motivated, for example to move into a larger house. Although this distinction is still used in migration studies, especially where direct information on the actual reasons for migration is not available, the idea has been questioned with the suggestion that the motivational distinction between short and long distance moves is not necessarily straightforward. Population geographers have argued that approaching the study of migration from the perspective of economic rationality, where decisions can be considered a result of an assessment of economic gains and losses, misses important elements of migration decision-making and experiences such as those to do with family and lifestyle choices at different stages of our lives (Bailey and Boyle, 2004; Bailey, 2009). Approaches taken by population geographers will be discussed more below.

## Timing of migration

Timing is a second dimension that can be used to classify migration. A move may be intended to be temporary or permanent. The timing definition used by the UN and many governments, including the UK, is that to be classed as international migration, a move should mean a change of residence for at least a year:

> A migrant into the UK is a person who has resided abroad for a year or more who states on arrival the intention to stay in the UK for a year or more, and vice versa for a migrant from the UK. (Home Office, 2006)

However, the frequency of short-term migration (for less than one year) has increased in recent years and agencies such as the Office for National Statistics (ONS) in England and Wales have been developing methods to estimate short-term international migration. The numbers of short-term migrants (and international migrants) to Britain have to be estimated because there are no direct measures or records of their arrival and departure. Measuring migration is considered further below. The ONS classifies short-term international migrants as those who visit England and Wales for more than one month but less than one year. Short-term migrants are thus categorised in between visitors and international migrants.

This leads us to consider the distinction between migration and travel or even commuting. If someone lives in a place for several weeks or months of a year are they a migrant? If someone lives in two or more places for roughly equal time, are they a migrant, and to which of these places? These questions are being raised as people's patterns of living and working are changing to become more mobile and more diverse in their mobilities.

A body of work focusing on mobilities is concerned with movement in everyday life, including for travel and commuting. Mobilities are distinct from, though related to, migration in that they are frequent movements which are part of everyday life. In contrast, migration is understood as a change of usual residence associated with some life change. The distinction between mobilities and migration is somewhat artificial and both have, for a long time, been recognised as important social phenomena (e.g. Beaujeu-Garnier, 1978; Clarke, 1972). However, as the broad field of mobilities studies has grown it has become helpful to draw distinction between migration and mobilities. This chapter deals with migration; for an overview of mobilities see Urry (2007), Merriman and Cresswell (2008) and Elliott and Urry (2010).

## Direction of migration

The direction of migration is linked to the temporality of migration. Migration may be a move from one place to another; a return from that place to the original place (return migration); onward movement to a third place (onward migration); or movement through a series or circuit of places. Return migration is, for example, a phenomenon that has been observed of immigrants from Poland who migrated to Britain after Polish accession to the European Union in 2004. Many have now returned and it has been suggested that this migration is following a pattern of short-term moves and then return (Burrell, 2009; Drinkwater et al., 2009).

Onward migration is, for example, an aspect of refugees' experiences. Refugees who have been given rights of residence and/or citizenship in a European Union country may choose to then migrate to another European Union country, for example to reunite with friends or family. Many refugees of Somali origin in the UK have migrated from other European countries such as the Netherlands because of the opportunities offered by the Somali communities in Britain (Lindley, 2007).

Circular migration has a long history in the form of pastoral nomadism. It also takes other forms including international migration of skilled workers (e.g. Findlay, 1988) and internal migration of seasonal agricultural workers. Hanson and Bell (2007) studied the seasonal migration of fruit and vegetable workers in Queensland, Australia. They suggest that although around half of the seasonal workers on the harvest trails are permanent itinerants, students also take on this work for additional income and retired people and working holidaymakers (from Australia and overseas) engage in seasonal agricultural work more through lifestyle choice than for economic reasons. This work demonstrates that one of the earliest forms of population movement endures but with the involvement of different people and places.

As well as thinking about migration in terms of types and patterns of movement, population researchers, especially geographers, are also interested in the specific places, the specific geographies, that are connected through migration. A complex combination of historical, political, economic, social, cultural and legal factors shape migration flows and means there are more connections between some places than others. For example, Britain's international migration links for most of the twentieth century were predominantly with countries of the Commonwealth, the former British empire; within the UK, London's economic status has a particular attraction for young adults seeking work (Parkinson et al., 2006).

## Reasons for migration

A fourth fundamental element of defining migration is the reason, or more likely, the reasons, for migration. A broad, and crucial, sub-division of reasons for migration is whether migration is forced or not. Forced migration is when people migrate against their will. Forced migrants are asylum seekers or refugees, pushed from their homes by fear of persecution or environmental disaster. People who have a well-founded fear of persecution have an international legal right to apply for protection (asylum) through residence in another country. There are two main routes for attaining refugee status: the first is to apply for asylum from within the country where protection is sought. These 'spontaneous asylum seekers' will then have their application assessed through the legal processes of that country. In some cases, this can involve being held in a detention centre or prison. The second route to refuge is to be recognised in the country of origin as a refugee by the UN and removed – usually from a refugee camp – to a country of safety. This route is becoming more common as it allows destination countries to have greater control over the entry of asylum seekers or refugees. Many more developed countries have an agreed annual quota of UN refugees who they will provide refuge for. For example, since 2004 the UK's Gateway Protection Programme has received up to 500 refugees a year through UN refugee agency resettlement. However, most refugees live in less developed countries, often within their countries of origin as Internally Displaced Persons. Box 6.2 presents some experiences of being a refugee.

Most migration is not forced migration: in 2010, 8 per cent of international migrants were refugees. Most migrants choose to move, usually for one of three primary purposes: work, study or family. Lifestyle choices also influence migration decisions. Although migration is a choice, it is still subject to constraints and opportunities which shape the migration processes and experiences differently for different individuals. Understanding the reasons that people migrate, or do not migrate, is an important component of migration studies within population geography. This theme is explored further in the next section.

## Box 6.2   Refugee experiences

"A few years ago I had to leave my country; it was really hard for me and my family. When the people broke into my house I was terrified – I thought something would happen to my family. They tried to shoot my dad but my sister started to scream. Because she was making lots of noise they shot her instead. They took my father away. I saw my own sister die. My family was like this: mum, dad, three sisters and a brother. But now it was me and my mum, my sister and brother. All of us buried my sister and left Afghanistan."

**Hannah, aged 11, Quoted in Amnesty International (1999: 23–4)**

"In Swansea, where I am living with other asylum seekers … we were both trying to prepare breakfast. I remember this guy was trying to fry an egg. I observed that he took an egg, broke it with a spoon, and then he put the egg-shell in the frying pan and threw the yolk and white into the rubbish bin. I said to him: 'What are you doing?'… he became aware of his mistake … I tried to cool him down. Finally he told me that he had got a refusal from the Home Office, and since then he was in constant fear. It is terrible. Isn't it?"

**'Claiming Asylum is World-wide' by Million Gashaw Woldemariam, In Charles et al. (2003: 88) Exile by Humberto Gatica**

I abandoned
my bones
in the uncertainty
of airports
I get lost
in cities
under the nightmares
of lugubrious hotels
some night
somebody dies
in my dreams
In others
I chase my way back
to the music
of my rains
and my broken
landscapes

**In Charles et al. (2003: 50)**

# Drivers of migration

Scholars of migration have attempted to understand migration first by describing migration patterns and how they vary between people and places. We will look at some of the patterns of migration below. Certain patterns can reflect certain flows, or types of migration, such as the circular seasonal agricultural migration described above. The causes, or drivers, of migration vary depending on the exact nature of the migration. Knowing more about the drivers of precise migration flows is the interest of much migration research.

One of the earliest studies to formally ask 'what drives migration?' was Ravenstein's 1885 analysis of population distribution in Britain and his later assessment of population movement in countries of continental Europe and North America (Ravenstein, 1889). Ravenstein's approach was economic: he aimed to understand the dynamics of demand and supply of labour or 'the mode in which the deficiency of hands in one part of the country is supplied from other parts where population is redundant' (Ravenstein, 1885: 198). From these studies, Ravenstein produced 'laws of migration' which are still referred to today (Box 6.3).

From Ravenstein's work we can identify several themes that still inform how we understand what makes migration happen. These are the idea that migration is a result of push and pull factors; the idea that migration is a result of individual and structural (societal) forces; and the idea that migration is selective.

---

## Box 6.3  Ravenstein's 'laws' of migration (1885: 198–9)

- Most migrants move a short distance; and the 'currents of migration' are towards centres of commerce and industry.

- The currents of migration form a stepwise cascade: those close to the towns move into them to be replaced by people from more remote areas and so on.

- The most remote areas see dispersion of their population to less remote areas in a stepwise process that is the inverse of movement to towns.

- Each main current of migration produces a compensating counter-current.

- Migrants proceeding long distances generally go by preference to one of the great centres of commerce or industry.

- The natives of towns are less migratory than those of the rural parts of the country.

- Females are more migratory than males.

## Push and pull factors

A migration event can be seen to be the result of a combination of factors which encourage an individual to leave a particular place with factors that attract the individual to another place. The assessment of the balance between these 'push and pull' factors is part of the decision-making processes which eventually result, or not, in the migration event. 'Push and pull' factors may combine work, housing, family, relationship, quality of life and place-characteristic considerations.

## Individual and structural factors

In migration decision-making, the factors to be taken into consideration are to do with the individual's circumstances such as their age or lifestage, their family situation, their housing situation, their economic situation; and also the societal structures which have an influence on the individual which may present opportunities or constraints for migration. Structural factors may include the legal framework for migration (particularly international migration), the labour market and the housing market.

Structural factors will interact with personal circumstances. For example Manuel Aalbers's (2005) work illustrates how the mortgage lending market can affect people's ability to move to particular places. His study of Rotterdam and Amsterdam in the Netherlands shows how mortgage lenders operated systems that restricted the granting of mortgage loans based on the residential address of the applicant. This process of 'redlining' is a form of place-based social exclusion which has been found in the US to particularly affect people living in areas of high ethnic minority population.

## Migrant selectivity

Since Ravenstein's work, migration scholars have demonstrated that migration is selective: some people are more likely to be migrants than others and some places are more likely than others to attract migrants. In other words, migrants are not representative of the general population. Understanding the ways in which migrants, and different types of migration, are selective is an important element of the study of migration. It may represent unequal opportunities and/or different aspirations.

Finney and Simpson (2008) investigate migrant selectivity by examining differences in levels of residential mobility within Britain between ethnic groups. They use census data to show that some minority ethnic groups have particularly low levels of residential mobility and that residential mobility differences between ethnic groups remain after other characteristics of the population known to be associated with migration, such as age, sex, tenure, socio-economic status and health, are taken into account.

In sum, migration happens as a result of numerous factors which encourage people to leave one place and go to live in another. We saw in the previous section that these factors can be to do with work, study, family or lifestyle, or a mixture of these. People's decisions about migration – how, where, when, who – take into account individual considerations and also societal factors, which also could be considered to be place-related factors. Together the various influences, operating at individual and societal levels, result in migration that is selective (or uneven) in terms of the people who move and the places they leave and go to.

## Measuring migration

The study of migration grew from approaches in economics, demography and population geography that were primarily concerned with numerical description – with counting migrants. As a result, a number of conventions for measuring migration were developed and these measures are still used to document migration patterns and trends. These measures include in-migration, out-migration, net migration, gross migration and migration intensity which are defined in Box 6.4. For further detail of migration measures see Rowland (2003) and Bell et al. (2002). Migration measures are often expressed as rates, that is, in relation to the size of the population of interest (such as the population of a city or country) which enables comparison of migration patterns between places. We will see some examples of migration levels for different countries in the next section.

Two important points of difference in the measurement of migration are, first, whether migration is measured as a stock or a flow; and second, whether migration is measured as an event or transition. Migrant stocks are the numbers of migrants in a particular place at a particular time; migrant flows are the numbers of moves between specified places over a defined time. Migration event data capture every move (in a defined area and time); migration transition data capture changes in residential circumstance over a defined time period.

As well as descriptive measures of migration, quantitative study also uses advanced statistical and mathematical modelling. Quantitative modelling aims to describe a system using mathematical language. The development of mathematical and statistical models to describe (and explain) migration is an area of ongoing research (see Raymer and Willekens, 2008).

Models take a number of forms depending on their conceptualisation of migration. One important dimension in this conceptualisation is whether migration is being measured, analysed and modelled for individuals or areas (places). This will depend on the research questions and on the data available. For example, the question 'what are the characteristics of migrants compared with non-migrants?' would require individual level analysis with information (data) about individuals. The question 'what was the impact of migration on the population size of Manchester in the 1990s?' would require area-level data and analysis (see Chapter 3 for discussion of population data and scales of analysis). Much geographical work is concerned with place, and data and methods (e.g. multi-level analysis) now exist to allow analysis that combines individuals and areas.

---

### Box 6.4    Measures of migration

For a given area over a given time period:

In-migration/immigration = number of arrivals

Out-migration/emigration = number of departures

Net migration = difference between the number of arrivals and departures

Gross migration = the sum of arrivals and departures

In/Out/Net/Gross migration are often calculated as rates i.e. in relation to the population of an area. This is usually expressed as a percentage or per 1,000 population. This allows comparison of places and times.

Migration effectiveness: the effectiveness of migration as a process of population redistribution. It summarises the extent to which migration of a place represents an exchange of population or a shift of population towards or away from the place.

Migration effectiveness ratio = net migration/gross migration × 100

---

An early approach was gravity modelling which was based on Newton's law of gravity and attempted to apply physics to population systems. The model is based on the idea that the greater the importance of a place by a given measure, the greater the in-movement to the place tempered by the distance of the places from each other (distance decay). Gravity modelling can be considered to be spatial analysis. Developments in this field are being pioneered by geographers making use of advances in data and geographical information systems. Lloyd (2010, 2011) provides an overview of models and their application.

Regression modelling is a common approach and is used to predict migration outcomes for individuals or locations based on information that is known about those individuals or places. Recent developments in migration modelling include probabilistic models, longitudinal approaches, multi-level modelling, spatial analysis and social network analysis. Probabilistic models estimate the probability of a (migration) event occurring (for a given place, population group, etc.) based on historical data. This approach is used in population projection and forecasting as well as in estimation of migration where full information is not available (see Raymer and Willekens, 2008).

Longitudinal approaches are dynamic in that they take time into account. For an example see Cooke et al. (2009). Event history analysis is an example of a longitudinal modelling approach that has been applied to migration. Multi-level modelling combines information from different levels or scales. For example, a migration event can be predicted using information about individuals and about places. Kulu and Billari (2004) take this approach in their study of internal migration in Estonia.

Social network analysis views a social system (or structure) as a collection of individuals (or organisations; nodes) tied to one or more others in the network by some form of interdependency. Conceptually, this approach is valuable to migration studies (see Vertovec, 2003) but methodologically application of social network analysis methods has been restricted by data availability.

In fact, data availability is a challenge for the (quantitative) study of migration generally. Information about migrants is difficult and costly to collect because migrants, particularly international migrants, are a minority of the population and they are difficult to locate because, by definition, they have moved from one place to another! We have learnt in Chapter 3 about data sources for population geography, including censuses, surveys, administrative data and population estimates. In Britain, for example, all of these sources provide some information about migration patterns but each source has its limitations. The census is the most comprehensive source of data because it covers the whole population. However, migration is defined on the basis of a change of address in the year prior to the census so not all migrants or migrations are captured. Also, the census in the UK only takes place every ten years. Survey and administrative sources may ask about people's migration but often the number of people in these datasets (the sample size) is relatively small which makes it difficult to conduct robust quantitative analysis, especially for subgroups of the population (such as particular age groups or ethnic groups).

As it has been recognised that the totality of a migration system cannot be captured and modelled, and there has been a shift towards humanist and postmodern approaches to migration studies, scholars have used qualitative research methods. Qualitative approaches include ethnographies, case studies, interviews, focus groups, participant observation, visual methods, participatory and action research. Qualitative methods are particularly helpful for investigating migration decision-making, migration motivations and experiences of migration which are not easily captured by quantitative methods.

# Migration over time and place

We saw at the start of this chapter some interesting statistics on migration. However, we have also seen that migration can be difficult to measure! It is quite a challenge, therefore, to compare migration between places and times and it is always important to consider the data sources and definitions being used. In this section, we will look at two sets of data to compare migration over time and place. First, we will look at UN data to compare international migration between countries. Second, we will look at UK census data to compare internal migration for subnational areas and population subgroups.

## Comparing countries' international migration using UN data

Figure 6.1 shows international migration (estimated) for the world and for more developed and less developed regions between 1990 and 2010. The rise in the number of

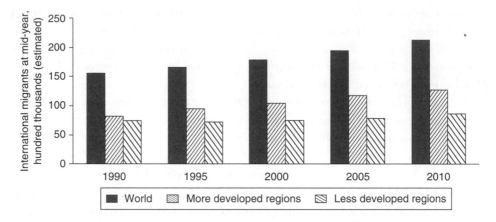

Figure 6.1   Estimated number of international migrants at mid-year, 1990 to 2010,
world regions

*Source:* UN Department of Economic and Social Affairs, Population Division (2009a)

international migrants globally over this time is immediately clear; and this was the case for more and less developed regions as well as the world. In 2010, there were estimated to be 214 million international migrants in the world, of which 128 million were in more developed regions and 86 million were in less developed regions.

We can examine the same data for individual countries. A selection of countries is shown in Table 6.1. The countries are listed in descending order of number of international migrants in 2010. Of these countries, the US has the highest number of international migrants through the 1990s and 2000s, with 43 million international migrants in 2010 (this is equivalent to around two thirds of the UK population). The UK ranks second in each year, with 6.5 million international migrants in 2010. Table 6.1 shows that the more developed countries tend to have larger stocks of international migrants than the less developed regions which is in line with Figure 6.1.

Most of the countries in Table 6.1 experience an increase in their international migrant stock between 1990 and 2010. However, the two South American countries, Mexico and Brazil, experience a dip in international migrants in the mid-1990s before the figure rises through the 2000s towards 1990 levels. Zimbabwe and Egypt's international migration stocks also fluctuate over the two decades.

An alternative way to compare international migration across countries is to think about international migrants as a proportion (a percentage) of the country's population. This measure is presented in Figure 6.2 and Table 6.2. Figure 6.2 shows that around 3 per cent of the world's population are international migrants. This figure has been stable over the 1990s and 2000s. We know that absolute numbers rose (Figure 6.1) so this tells

Table 6.1    Estimated number of international migrants at mid-year, 1990 to 2010,
             selected countries

| Country | 1990 | 1995 | 2000 | 2005 | 2010 |
| --- | --- | --- | --- | --- | --- |
| United States of America | 23,251,026 | 28,522,111 | 34,814,053 | 39,266,451 | 42,813,281 |
| United Kingdom | 3,716,271 | 4,190,617 | 4,789,678 | 5,837,750 | 6,451,711 |
| Spain | 829,705 | 1,041,191 | 1,752,869 | 4,607,936 | 6,377,524 |
| Australia | 3,581,363 | 3,853,736 | 4,027,478 | 4,335,846 | 4,711,490 |
| Japan | 1,075,626 | 1,362,512 | 1,686,567 | 1,998,884 | 2,176,219 |
| Sweden | 777,571 | 905,628 | 992,623 | 1,112,917 | 1,306,020 |
| Mexico | 701,088 | 457,837 | 520,725 | 604,670 | 725,684 |
| Brazil | 798,517 | 730,517 | 684,596 | 686,309 | 688,026 |
| China | 376,361 | 437,269 | 508,034 | 590,252 | 685,775 |
| Zimbabwe | 627,098 | 432,504 | 411,410 | 391,345 | 372,258 |
| Egypt | 175,574 | 174,301 | 169,149 | 246,745 | 244,714 |

*Source:* United Nations Department of Economic and Social Affairs, Population Division (2009a)

us that numbers of international migrants are growing in line with growth in global
population. However, the patterns differ for more developed and less developed
regions. International migrants were, in the 1990s, around 7 per cent of the population
of more developed regions. This figure rose steadily over the next two decades to just
over 10 per cent. In contrast, migrants as a percentage of the population of less developed regions fell slightly from just below 2 per cent in 1990.

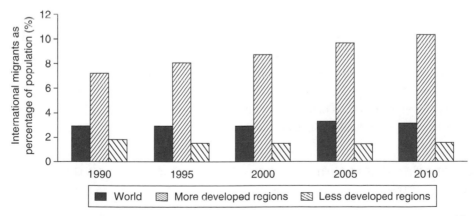

Figure 6.2    International migrants as a percentage of the population, 1990–2010, world
              regions

*Source:* UN Department of Economic and Social Affairs, Population Division (2009a)

Table 6.2   International migrants as a percentage of the population, 1990–2010,
            selected countries

| Country | 1990 | 1995 | 2000 | 2005 | 2010 |
|---|---|---|---|---|---|
| Australia | 21.0 | 21.3 | 21.0 | 21.3 | 21.9 |
| Spain | 2.1 | 2.6 | 4.4 | 10.7 | 14.1 |
| Sweden | 9.1 | 10.3 | 11.2 | 12.3 | 14.1 |
| United States of America | 9.1 | 10.5 | 12.1 | 13.0 | 13.5 |
| United Kingdom | 6.5 | 7.2 | 8.1 | 9.7 | 10.4 |
| Zimbabwe | 6.0 | 3.7 | 3.3 | 3.1 | 2.9 |
| Chile | 0.8 | 0.9 | 1.2 | 1.4 | 1.9 |
| Japan | 0.9 | 1.1 | 1.3 | 1.6 | 1.7 |
| Mexico | 0.8 | 0.5 | 0.5 | 0.6 | 0.7 |
| Brazil | 0.5 | 0.5 | 0.4 | 0.4 | 0.4 |
| Egypt | 0.3 | 0.3 | 0.2 | 0.3 | 0.3 |
| China | 0.0 | 0.0 | 0.0 | 0.0 | 0.1 |

*Source:* United Nations Department of Economic and Social Affairs, Population Division (2009a)

   Table 6.2 presents international migrants as a percentage of the population for 1990
to 2010 for the same selected countries shown in Table 6.1, again ordered from highest
to lowest on the 2010 figure. The order has changed from that in Table 6.1. Australia
now tops the table with around 22 per cent of its population being international
migrants. In comparison, the figure is 0.1 per cent for China and 0.3 per cent for Egypt.
The US, which was top in terms of numbers of migrants, ranks fourth amongst these
selected countries with 13 per cent of the population being international migrants. The
UK comes mid-table: 10 per cent of the population were international migrants in
2010, an increase from the figure of 6.5 per cent in 1990. Most countries in Table 6.2
see a rise in the proportion of the population that are international migrants. However,
this is not the case for Brazil, Egypt and China, or Zimbabwe which experienced falls
from 6 to 3 per cent over the two decades. The greatest rise between 1990 and 2010 is
for Spain: 2 per cent of its population were international migrants in 1990; by 2010 the
figure was 14 per cent. The expansion of the European Union in the 2000s is an important factor in Spain's international migration experience as well as ongoing migration
flows between Spain and Latin America.

## Comparing internal migration for areas of Britain using census data

The UK census includes a question on place of usual residence one year prior to census
day. This allows people's residence at the time of the census and one year previously to
be compared. If a person changes address, they are considered to be an internal migrant.

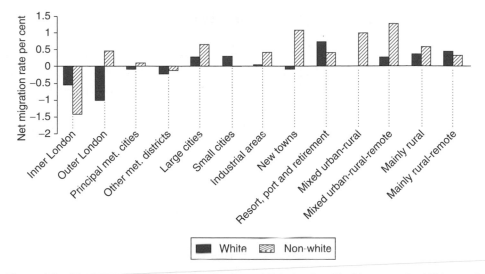

Figure 6.3   Net internal migration for districts of Britain classified by type, for White and non-White population, 2000–2001

*Source:* Simpson and Finney (2009) using 2001 Census Special Migration Statistics

These data also allow calculation of the moves into and out of subnational areas. Data which give information about migration flows between areas are called 'interaction data'. The census is the best source of interaction data for small areas in the UK.

Figure 6.3 uses interaction data from the 2001 Census to compare net migration for districts. Districts are administrative areas with an average population of 130,000. There are 378 districts in Britain. The districts have been categorised according to their urban-ness (we will see in the next section that urban-ness is an important concept for studying migration). The populations of the districts have been categorised by ethnic group into 'White' and 'non-White'. Migration is measured in Figure 6.3 as net migration rate. That is, the balance of in-migration and out-migration for each group of districts expressed as a percentage of the population of that group of districts.

The leftmost bars in Figure 6.3 show the net migration rate for Inner London. Both White and non-White population left London on balance (more people left than moved in) between 2000 and 2001. Non-White population migrated away from London at a faster rate than White population. The rightmost bars in Figure 6.3 show the net migration rate for 'mainly rural or remote' districts. These types of districts gained White and non-White population on balance through internal migration between 2000 and 2001. The pattern for the bars in between the two extremes shows a general shift from net out-migration to net in-migration with increasing rural-ness. This tells us that Britain's population is suburbanising and counterurbanising and that this is the case for the White and non-White populations. It is worth noting that if the graph is produced using 1991 Census data the pattern is almost identical (see Simpson and Finney, 2009).

# Approaches to studying migration

Understanding what makes migration happen – how push and pull factors interact; the meaning of individual and structural factors; how and why migrants are self-selected; and what explains the patterns of migration we have just seen – as well as understanding the meaning of migration for individuals and places, can be approached from a number of conceptual standpoints. The conceptual standpoint will affect the methodological approach to studying migration and the nature of the findings. Boyle, Halfacree and Robinson (1998: 57–58) divide conceptual approaches into determinist, humanist and integrated, which they define as follows:

> Determinist approaches 'play down the role of the individual in actively deciding whether or not to migrate by assuming migration to be an almost inevitable response to some rational situation ... humanist approaches stress the importance of seeing the individual migrant as an active decision-maker, whose ultimate decision to migrate may or may not be rational from any particular perspective'. Integrated approaches 'argue that a balance needs to be struck between regarding migration as an inevitable response to particular circumstances and seeing it as a completely individual action'.

Early studies of migration, including Ravenstein's, were deterministic, attempting to find rules, or laws, that could be universally applied to migration. This approach is still common in some fields. For example, economists approach migration as an economic act which is primarily determined by economic circumstance and economic risk (potential losses and gains). However, there has been a shift in migration studies more broadly to an integrated approach and, over the last three decades, there have been deliberate efforts to develop integrated migration theory. A central aim of this theoretical development has been to enable explanation, as well as description, of migration patterns and processes.

There are many approaches, or perspectives, taken by population geographers who want to understand migration patterns and processes. Here we will review three perspectives which are important in migration research: urban-ness/rural-ness; families and life course; and ethnic integration. These examples are illustrative; you can find out more in the recommended readings at the end of this chapter.

## An urban-rural perspective

As we have seen from Ravenstein's studies of migration in the 1880s, the concepts of urban and rural, town and country have always been important for migration studies. As outlined in Chapter 2, in the 1970s Zelinsky (1971) proposed a Hypothesis of Mobility Transition by combining the 'laws of migration' with the Demographic Transition Model. In the five phases of societal advancement that Zelinsky identifies taking place alongside demographic transition, the first three are characterised by 'countryside to city' migration; in the fourth there is the added component of city to city moves; in the fifth phase 'a future superadvanced society', Zelinsky hypothesises interurban, intraurban and circulatory moves to dominate.

The accuracy of Zelinsky's ideas about the direction and level of migration between urban and rural areas can be debated. Whilst less developed nations are still experiencing urbanisation due to migration, a trend somewhat different from Zelinsky's hypothesis has been identified for more developed countries. This trend is counterurbanisation.

The concept of counterurbanisation has been alive in migration research since the 1970s and the process has been shown to be an experience common to many more developed countries (Champion, 1989). Early work theorised counterurbanisation as a lifestyle migration of middle-class urbanites aspiring to a rural family life. More recently, researchers have suggested that counterurbanisation embraces a broader range of people and experiences (Halfacree, 2008).

The urban-ness or rural-ness of a place is thought to be important for migration decisions and experiences because of the general differences between urban and rural areas in terms of greenspace, housing markets and job markets. Examining the significance of these characteristics at different times and places continues to be a focus of migration research in population geography.

## A families and life course perspective

Migration research, being influenced as it has been by economic approaches, has traditionally focused on the experiences of individuals, particularly men (seen as the 'breadwinners'). However, most migration involves families or households, though not necessarily moving as one unit. Economic logic approaches theorised that migration decisions would be based on job or career opportunities of the male 'head of household' and that (female) partners would follow the move as a 'tied migrant' or a 'trailing spouse'. Since the 1990s, scholars have questioned these theories and their assumptions that migration decisions are primarily made on the basis of economic loss or gain (Bailey and Boyle, 2004). Scholars are examining migration as a family event, interrogating the social relations within families and households that contribute to migration decisions and experiences. The perspective of all members of families, including children, is thought to be important for understanding migration (Bushin, 2009).

Two themes in broader social science thinking have influenced this direction in migration studies. These are a life course approach (Bailey, 2009) and the recognition of a Second Demographic Transition (Van de Kaa, 1999; see Chapter 8). It has been long recognised that migration patterns vary by age and the age profile of migration shown in Figure 6.4 (for Britain) is remarkably stable across time and place: migration rates are low in childhood and late adulthood, rise slightly in older age and have a striking peak in young adulthood. However, it is relatively recently that this age-differentiation in migration has been thought of in terms of processes of transitions through life stages (for example, see *Demographic Research* (2007), *Population, Space and Place* (2008) and Wingens et al. (2011)).

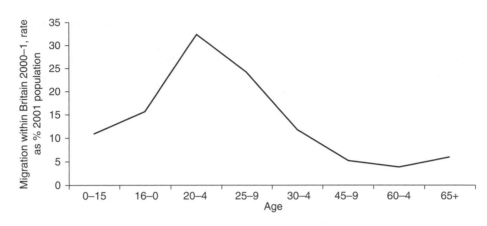

Figure 6.4    Internal migration rates (for Britain) by age

*Source:* 2001 Census Sample of Anonymised Records. Demoninator is 2001 population in each age group

Another important influence on migration study has been re-thinking of the meaning of 'family'. One characteristic of the Second Demographic Transition (which we discuss in Chapter 8) is its recognition of increased and increasing diversity of family and household compositions and relationships. The changing meaning of family matters for understanding migration because 'family' impacts on migration decisions and migration experiences, and family events (e.g. union, union dissolution, having children) are often associated with migration (Flowerdew and Al-Hamad, 2004).

## An immigrant/ethnic integration perspective

A longstanding theme of migration research has been its consequences in terms of changing the ethnic composition of a place. It is international migrants – Pakistanis in Britain, Hispanics in the US, Romanians in Spain – who become ethnic minorities. The issue of immigrant/ethnic integration is important for individuals' life experiences and for policy that aims for equality and social cohesion. For example, many Western European and North American countries have adopted area-based initiatives to address issues of economic polarisation. These initiatives, implicitly and explicitly, are ethnicised because minorities live disproportionately in the poorest neighbourhoods.

Ethnic integration is studied from many disciplinary perspectives and it has been long recognised that there are many dimensions to integration (Gordon, 1964). A particular interest of population geographers is the residential integration of minorities within countries and how this relates to other (social) dimensions of integration. In the mid-1990s, questions were asked about whether there were ethnic 'ghettos' (e.g. Peach, 1996a) and 'balkanisation' (Frey, 1995); and discussions were had about whether residential segregation is 'good' or 'bad' (Peach, 1996b) and how this changes over immigrant generations (for the children of immigrants) (Portes and Zhou, 1993).

These debates were renewed in the mid-2000s in the context of concerns in Europe and North America about international terrorism and religious fundamentalism (see Kalra and Kapoor, 2009). Population geographers in Britain have argued that claims of increasing ethnic residential segregation and resulting cultural conflict are misplaced; rather, there is increased ethnic mixing residentially (and otherwise) and common patterns and aspirations of residential mobility across ethnic groups (Finney and Simpson, 2009; Peach, 2009). Population projections using innovations in methods of population estimation predict that Britain's population will become more ethnically diverse and more mixed (Rees et al., 2011).

## Summary

- Migration is important as the third component of population change and the one that affects the distribution – the geography – of the population. Migration is also a significant event in people's lives and shifts in migration patterns reflect social change.

- Although migration can be simply understood as a change of residence, it can be quite difficult to specify a precise definition. Four important elements that define migration are boundary crossing (or distance), timing of migration, direction of migration and reasons for migration.

- Migration is the result of numerous interacting factors. Three themes of research that try to understand what makes migration happen are push and pull factors, the role of individual and societal forces and migration being selective.

- Migration is measured quantitatively in censuses, population registers and surveys. Standard measures are used to describe migration patterns and statistical modelling is used for further description and explanation of patterns. We have seen examples of migration patterns across countries and over time using UN data and within Britain using census data. Qualitative methods also add considerably to our understanding of migration.

- We have also discussed three approaches to understanding migration that are taken by population geographers: an urban-rural perspective; a families' and life course perspective; and a immigrant/ethnic integration perspective.

## Recommended reading

General texts on migration include:

Boyle, P., Halfacree, K. and Robinson, V. (1998) *Exploring Contemporary Migration*. Harlow: Longman.
Castles, S. and Miller, D. (2009) *The Age of Migration*, 4th edn. Basingstoke: Palgrave. This book includes guides to further reading.

Koser, K. (2007) *International Migration. A Very Short Introduction*. Oxford: Oxford University Press. This book includes a useful annotated bibliography.

Samers, M. (2010) *Key Ideas in Geography: Migration*. Abingdon: Routledge.

Texts of quantitative analysis of internal migration:

Champion, A.G. and Fielding, A. (eds) (1993) *Migration Processes and Patterns: Research Progress and Prospects*. Chichester: John Wiley and Sons.

Stillwell, J., Rees, P. and Boden, P. (eds) (1991) *Migration Processes and Patterns: Population Redistribution in the UK*. Chichester: John Wiley and Sons.

Migration theory (focus on international migration):

Brettell, C. and Hollifield, J.F. (eds) (2008) *Migration Theory. Talking Across Disciplines,* 2nd edn. Abingdon: Routledge. This includes a chapter on geography by Susan Hardwick.

Measuring migration:

Rowland, D. (2003) *Demographic Methods and Concepts*. Oxford: Oxford University Press.

## Organisations that provide information on migration:

Council of Europe
European Migration Network
Eurostat
Information Centre about Asylum and Refugees (UK)
International Organisation of Migration (IOM)
Metropolis
Mexican Migration Project
OECD
Refugee Council
UN
World Bank

## Organisations providing data and data support in the UK

National statistical agencies: ONS (England and Wales), GROS (Scotland), NISRA (Northern Ireland)
SOPEMI reports
UK Data Archive
UK Data Service

# 7

# LIVING ARRANGEMENTS

To investigate the relationship between population and societies we need to take account of the organisation of living arrangements and the ways that individuals interact on a daily basis. While the individual may be taken as the basic unit of analysis for the study of population, to fully appreciate the complexity of demographic events, it is important to consider how these events are experienced, not by individuals in isolation, but by groups of people living together in household units and/or as members of family or kinship groups. This chapter explores the living arrangements of different societies over space and time. The questions and issues addressed in this chapter are:

- How might we define a family and a household as the basic units of living arrangements?

- How does household composition vary over time and place? In what ways are changes in household composition associated with changing functions of the household unit?

- How many households do women head? What factors might account for recent increases in female-headed households and lone-parent families?

- What are the characteristics of one-person households and how can we account for recent increases in the number of people living alone?

In most societies, individuals live together in relatively small domestic groups. Most of the organisation of the day-to-day requirements of providing shelter, feeding and caring for individuals is carried out within these co-resident household units. Household units formed around family groups, or groups of related individuals, are by far the most common approach to living arrangements. However, this approach to living arrangements is not universal. We can think of more community based approaches to living arrangements, such as kibbutzim. Furthermore, household units need not necessarily be based around family units. For example the Ashanti of southern Ghana adopt a variety of living arrangements: some spouses live together, while other married couples live apart, with some married men living with their sisters and married women with their mothers (see Verdon, 1998). In North America and Northern Europe, more couples are forming 'living apart together' relationships, and maintain independent households rather than living together (Haskey, 2005; Levin, 2004). However, despite these exceptions, in most societies in both the developing and developed world, family-based households dominate the organisation of living arrangements.

In this chapter, we shall consider the main historical and contemporary trends in living arrangements in both developing and developed world. While changing patterns of living arrangements have received less attention than other demographic processes, as we explore in this chapter they have implications for resources, care and intimacy. In particular the trend towards more individualised living arrangements in the developed world not only increases demand on the housing stock but is also indicative of new forms of intimacy and the provision of care within relationships and family, and that co-residence is not a requirement for these activities. We shall return to this theme at the end of the chapter, but we begin by considering how we might define households and families.

However, before we consider definitions of households and families, consider the following information about families; by the end of the chapter you should be able to make sense of these data (all taken from US Census Bureau historical database):

- In 1900, the average household size in the US was 4.60; by the year 2000 it stood at 2.59.

- In 1900, there were just over 15 million households in the US; by 2000 the number of households had increased by over 600 per cent.

- Between 1940 and 2000, the number of female-headed family households in the US increased by just under 400 per cent.

- In 1940, 7.7 per cent of all US households were one-person households, by 2000 this had increased to 25.8 per cent.

## Definitions of households and families

Families and household arrangements are so much part of our everyday experiences that we rarely stop to think about how we might define them. Yet if we are to document and describe living arrangements, as well as understand the impact that living arrangements have on everyday life, then we need to be able to define what we mean by families and households.

# Family

Let us start with thinking about the family. A useful way to approach this is to think about what family means to you. Maybe you could think about who is in your family. Your parents, siblings, grandparents, uncles, aunts and cousins, maybe children of your own or a partner. But what about family friends? You might well know people whom you call aunt or uncle who are not actually related to you. You might also want to include family pets as part of your family. What is important is that there is no universal definition of family; your own interpretation of family might be different from that of your friends', or even members of your own family. You will probably use varying definitions of who is a member of your family for different purposes. For example having 'the family' around for Christmas may suggest a more inclusive definition than more everyday uses that map more closely to the nuclear family group.

However, for the purposes of government policy and the collection of survey data about families, this subjective approach to defining families is not always appropriate. In particular, in order to summarise the characteristics of family formation in populations, it is necessary to use a more rigid definition of the co-resident family group. The most common approach to classifying the distribution of family types in a population is to restrict family membership to the nuclear family (Ermisch and Overton, 1985; Haskey, 1996), where a nuclear family group is defined as:

> **either** a married, civil partnership or a cohabiting couple with/without unmarried/unpartnered children;
>
> **or** a lone parent living with unmarried/unpartnered child(ren).

It is also common to distinguish between families with dependent or non-dependent children. In the UK, the Office of National Statistics defines a dependent child as being aged less than 16, or aged between 16 and 18 and in full-time education, excluding all children who have a spouse, partner or child living in the household (ONS, 2007). Individuals who are neither living as a couple, or as parents or children are not identified as living in a nuclear family group; this will include people living on their own, with non-relatives or with relatives outside their immediate nuclear family. This approach to defining family membership does not therefore necessarily correspond to whom we might regard as members of our family. In particular it excludes other relatives such as grandparents, uncles/aunts. For example, a grandparent living with a married daughter, her partner and children would not be classified as a member of the daughter's nuclear family unit, but would be classified as a 'non-family' person. A lone mother living with her daughter and her parents would be classified in a distinct family group (a lone parent family group) from her parents (a couple without children). Various combinations of nuclear family groups living in the same household are illustrated in Figure 7.1.

Figure 7.2 illustrates how this definition of nuclear family membership can be applied to classifying family groups in populations, and gives the breakdown of family types for the UK using the 2001 Census. We can see that over two-thirds of all families consist of a married couple with or without (non)dependent children, though there are slightly more married couples with (non)dependent children. Approximately one in six families are lone parent families, while cohabiting couple families account for the remaining 13.2 per cent; in contrast to married couple families, there are more cohabiting couple families without children than those with (non)dependent children.

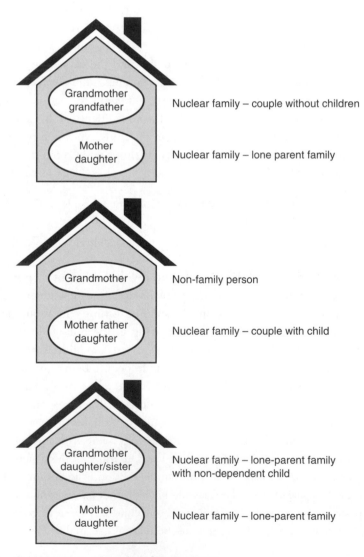

Figure 7.1    Combinations of nuclear family groups

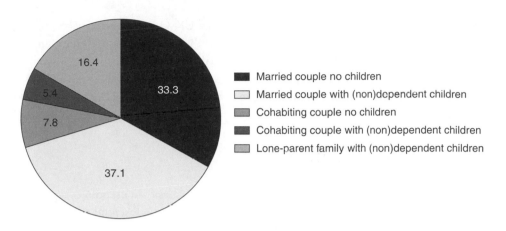

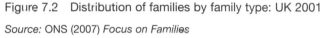

Figure 7.2    Distribution of families by family type: UK 2001

*Source:* ONS (2007) *Focus on Families*

## Households

Households are the basic units of living arrangements in most societies, usually formed around the nuclear family group. The household is used for the purposes of data collection, for example, censuses are almost always carried out by enumerating individuals within household units. For this reason there is far more uniformity in approaches to defining households. For example, in the 2011 UK Census a household was defined as follows:

- one person living alone; or

- a group of people (not necessarily related) living at the same address who share cooking facilities and share a living room or sitting room or dining area.

However straightforward this definition may appear, there are always problems in defining households. The first problem to consider is defining the boundary of the households. Students' experiences are appropriate here. For example, when does a shared house of six individuals become a household of six people? It is not possible to provide a precise answer here; some house-shares will very much conform to the accepted understanding of a household, pooling resources and sharing tasks between household members. Other groups may choose very separate living arrangements. Moreover, which household individuals belong to is not always apparent. Student households again can be problematic. If a student is living away from home during term time, is she also a member of her parents' household? Again there is not a definitive answer to this problem. However, for the purposes of data collection, guidelines have to be drawn up which deal with these issues, to allow for consistency in locating individuals in households.

The challenges of consistently defining household membership are certainly not diminishing, rather statistical agencies are increasing aware of the difficulty in assuming that people will have a unique 'usual' residence. As a UK Office of National Statistics' review of proposals for the 2011 Census identified: 'increased mobility and complex living arrangements mean that the concept of usual residence is becoming less clear. There are significant numbers of people who no longer associate themselves with a single residential address, thus there has been a weakening of the link between people and dwellings' (ONS no date: 4).

## Household composition

Once a household has been defined, it is then possible to classify household composition for a population. By far the most common way of classifying household types is based on the combination of family units in a household. There are two important distinctions of household types that are used to classify household composition:

*Family versus non-family households.* While most households contain family units, this is not always the case, such as people living alone or friends sharing. Hence, a useful distinction to make is between family and non-family households. The former can then be further characterised according to family type, while the latter can be distinguished between households consisting of related or unrelated individuals.

*Simple-family households versus extended, complex or stem households.* This distinction applies to family households and essentially distinguishes households consisting of one nuclear family group only, and those households which either contain one nuclear family and other related or non-related individuals, or two or more nuclear family groups, as follows:

*Simple family households*: one nuclear family only; all household members are part of the same nuclear family group.

*Extended family households*: nuclear family with other related or non-related individuals.

e.g.　Mother, father, children and grandmother

Couple and female partner's sister

Mother, children and family friend

*Complex or multi-family households*: two or more nuclear family groups

e.g.　Mother, father, daughter and granddaughter

e.g.　Mother, father, children, grandmother and grandfather

Extended and complex households are sometimes described as 'stem' family households. The term stem family was first defined by the French sociologist Frédéric le Play in 1891 (McIssac Cooper, 1999) and refers to a particular pattern of inheritance, where one child (usually the eldest boy but not necessarily) inherits the family property and remains living with parents, while other children leave to seek employment elsewhere.

While the term stem family is often used in analysis of household composition, it conveys more than just an organisational structure, but hints at the wider functions of the household. Examples of extended and complex households that are also stem family households include:

Mother, father, children and grandmother

Mother, father, children, grandmother and grandfather

## Characteristics of households

Comparing the characteristics of households in different societies is not just a question of classifying who lives with whom, but can shed light on the ways in which individuals share common activities, such as production and consumption (of food and other material goods) as well as looking after children, the sick and elderly. For example, in societies without, or at best a rudimentary, welfare state support for the elderly or the sick, provision of care may be most efficiently arranged within a co-resident household group. In pre-industrial societies where food production was carried out within domestic groups, the organisation of household structure might be based on larger and more complex units to maximise household resources. Information on the size and complexity of household groups can shed some light on the different functions and forms of households in both historical and contemporary societies.

## Household size

It might be assumed that the most straightforward household characteristic to compare is household size. Table 7.1 gives average household size for different countries for the last available census year. We can see that there is clear distinction between smaller average household sizes in developed countries (Japan, Italy, the UK, the US) compared to the larger average size recorded for developing societies (the Gambia, Pakistan, India and Ghana). However, it should be noted that household size is not routinely collected by the major non-government organisations and the reliability of this data is questionable.

Notwithstanding the problems with the data, we might anticipate that larger average household sizes in developing countries are associated with higher fertility and/or a greater propensity to form extended or complex family households. In other words, households are larger in countries such as the Gambia and Pakistan either because households in these countries, on average, include more children (higher fertility) and/ or because there are other relatives or non-relatives living together in the same household. Conversely, smaller average households in developed countries are associated with lower fertility and simple nuclear-family household formation.

Table 7.1   Average household size for select countries, various years

| Country | Year | Average household size |
| --- | --- | --- |
| Gambia | 1993 | 8.9 |
| Pakistan | 1998 | 6.8 |
| India | 2001 | 5.3 |
| Ghana | 2000 | 5.1 |
| Malaysia | 2000 | 4.5 |
| Mexico | 2000 | 4.4 |
| Botswana | 2000 | 4.2 |
| Australia | 2001 | 3.8 |
| China | 2001 | 3.4 |
| Italy | 2000 | 2.8 |
| Japan | 2000 | 2.7 |
| USA | 2001 | 2.7 |
| UK | 2001 | 2.4 |
| Germany | 2000 | 2.3 |
| Sweden | 2000 | 2.1 |

*Source:* World Bank (2006) *World Development Indicators* Washington: World Bank, Table 3.11s

## Trends in household formation over time

One simple hypothesis states that economic modernisation is associated with the formation of nuclear family households. Hence we might expect to find that households in pre-industrial societies were very different from those in contemporary more developed countries and that changes in household composition over time reflect changes in the functions of the household. However, the historical record for some countries suggests more continuity in household size and composition over time than is often popularly portrayed. As the family historian Peter Laslett comments for English society:

> The wish to believe in the large, extended, kin-enfolding, multi-generational, welfare-and support-providing household in the world we have lost seems to be exceedingly difficult to expose to critical evaluation. (Laslett, 1965: 92)

Put simply, the popular notion that households in the past were large, and provided care for family members at all ages, is very strong and often is used in contrast to the smaller households found in contemporary industrialised societies. Yet as Laslett's quote suggests, this ideal may be little more than a myth in some contexts, and the ideal of the large extended household is one that has been subject to considerable academic debate.

---

## Box 7.1   Family reconstitution

This involves the reconstruction of the genealogical history of families using church records, i.e. baptism, marriage and death registers. The technique was originally developed by the French demographer, Louis Henry. The Cambridge Group of the History of Population have used the technique to reconstruct the organisational structure of families and households, kinship networks, inheritance systems, and migration patterns.

---

Table 7.2   Laslett's analysis of household composition for 64 English settlements, 1622–1854

|  | 30 most reliably recorded communities | 35 next most reliably recorded communities |
|---|---|---|
| Percentage of one-person households | 8.5 | 8.7 |
| Percentage of households with no family | 3.6 | 3.2 |
| Percentage of simple family households | 72.1 | 71.9 |
| Percentage of extended family households | 10.9 | 11.9 |
| Percentage of multiple family households | 4.1 | 4.1 |
| Percentage of households with: | 25.1 | |
| 1 generation | 69.2 | |
| 2 generations | | |
| 3 generations | 5.7 | |

*Source:* Laslett (1965: 98)

The study of household and family formation in England is most closely associated with Peter Laslett's research and his associates at the Cambridge Group for the Study of Population History. Laslett concluded that the evidence from family reconstitution for England shows that nuclear family households dominated from pre-industrial times onwards. There are two important indicators that Laslett used to support this. First, average household size for England has historically been relatively small and stable. From 1600 to the early twentieth century it has averaged around 4.75, smaller than that recorded for many contemporary developing societies. Second, the evidence from family reconstitution studies illustrate that the majority of households in pre-industrial and industrial England consisted of two generations only. A summary of the family reconstitution for the years 1622–1854 co-ordinated by Laslett is given in Table 7.2. We can see that over this period, at any one time 7 out of every 10 households consisted of simple nuclear family households, with 15 per cent comprising extended or multiple family groups.

Laslett therefore concludes that the belief that we lived in large extended households in the past is not backed up by the historical data and in fact the reality of family life in historical times was not that distinct from contemporary English societies. His approach to household and family in past times is not though without its detractors. There are several criticisms that have been made of Laslett's conclusions about households in the past (Anderson, 1995):

1   The first criticism relates to the reliability of data that Laslett's conclusions are based on. On average only about one-third of a population can be reconstituted, which suggests that there might be a problem of selection bias in the results. Put simply, it is easier to reconstitute simple households than more complex ones and this might inflate the proportion of simple nuclear households in the reconstituted population.

2   England is atypical and we should we wary in making generalisations to other parts of Britain and Europe. In particular historical evidence from Southern Europe, also southern parts of France and Germany suggests that extended households were more common.

3   There is a problem in using the mean household size as an indicator of household complexity. While a mean household size below 5 with 70 per cent of households containing no non-conjugal kin may appear as strong evidence of the absence of extended family households, we need to take into account the changing constitution of households over the lifecycle, and the opportunities to form extended households. An extended family household organisation may well exist, without there being much evidence of this type of household being present. The opportunity to form extended family households will depend on the timing of fertility and level of mortality. In contemporary societies where mortality levels are very low, it is usual for there to be at least three generations alive within a family at any one time, hence there are continual 'opportunities' to form an extended family household. Yet in historical times when life expectancy was much lower, the possibility of forming vertical extended family households was more limited. This raises an important question regarding the link between the households that family groups actually lived in and 'ideal' household types.

4   The final criticism to consider is that we need to account for diversity between different social groups. Laslett provides averages for communities, yet we might expect differences by social-economic group, e.g. between gentry, farmers and labourers. There might also be selection bias with evidence more readily available for some social groups given varying literacy levels and burial costs, etc. Some groups may thus be under-recorded and less available for family reconstitution.

In response, Laslett argued that he was not setting out to demonstrate that all households were nuclear family households, and that family groups would respond to particular circum-stances. Household composition was flexible and fluid with households changing from sim-ple forms to extended or multiple family households over time. On balance what is at issue here is the attempt to identify ideal types of living arrangements that are characteristic of societies at particular times. Ruggles's (2009) recent review concludes that the argument for distinctive patterns of living arrangement in historical Northern Europe and North America cannot be sustained. He argues that the contribution of cultural practices to living arrangements is more muted than historians such as Laslett have suggested, and that economic conditions have played a greater role in influencing inter-generational living arrangements.

## Contemporary household change

Despite Laslett's claim that English households in the past were closer in form to contemporary households than is often believed, it is too simplistic to assume that household characteristics have remained unchanged in more recent times. For example, Haskey's (1996) analysis of household size in Great Britain demonstrates that household size declined steadily throughout the twentieth century, from an average of 4 in 1900 to around 2.5 in the 1990s. Moreover, as the population has increased, the number of households has also increased, as people tend to live in smaller domestic groups compared to 100 years ago. Between 1971 and 2001, data from the Labour Force Survey for Great Britain shows that while the population increased by 5 per cent, the number of households increased by 31 per cent. These trends are replicated elsewhere, for example, data on household composition provided by the US Census Bureau also shows a tendency towards more people in smaller households. Data for total population, the number of households, family households and one-person households and household size for the United States from 1940 to 2010 are illustrated in Figure 7.3. The decline of household size, concomitant with an increase in total population is evident for the US, leading to more households and more one-person households. Between 1970 and 2003, the proportion of American households with more than five residents decreased from 21 per cent to 10 per cent, while at the same time the proportion of households with one or two members increased from 46 per cent to 60 per cent (Fields, 2003).

In some societies, the impact of transformations in living arrangements has been more pronounced. In Japan for example, data for the period 1980 to 2005 illustrate an important change in household composition (see Figure 7.4). Unlike England, traditional Japanese society has been associated with extended household formation (Budak et al., 1996). Yet, as is illustrated in Figure 7.4 the proportion of extended/complex households has declined from 20 per cent in 1980 to 12 per cent in 2005. Most of this decline in more complex households has been offset by an increase in one-person households, while the proportion of simple nuclear family households has remained relatively stable. To explore this change in more detail, it is germane to consider the living arrangements of elderly persons (Figure 7.5). Here again we see an important change. In 1980 the majority of elderly in Japan lived in extended family households; in

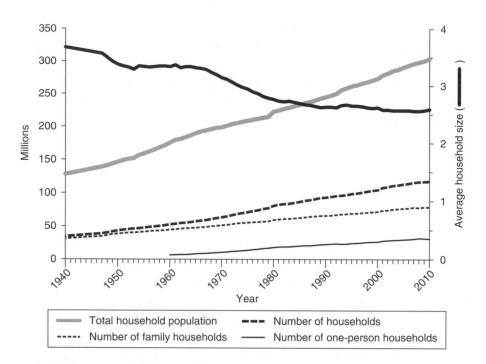

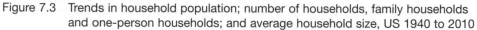

Figure 7.3    Trends in household population; number of households, family households
              and one-person households; and average household size, US 1940 to 2010

*Source:* US Census Bureau: Family and living arrangements: www.census.gov/population/www/
socdemo/hh-fam.html

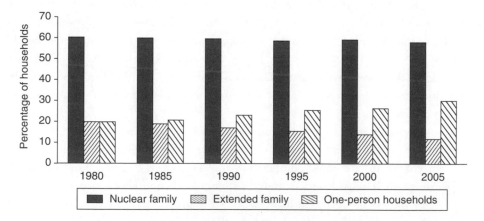

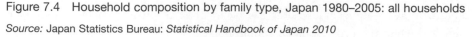

Figure 7.4    Household composition by family type, Japan 1980–2005: all households

*Source:* Japan Statistics Bureau: *Statistical Handbook of Japan 2010*

many cases this would involve living with a married child (usually a son). Yet by 2005
the most common living arrangement for elderly Japanese is in a nuclear family; that is
living with a spouse. This trend will in part reflect increases in Japanese life expectancy,

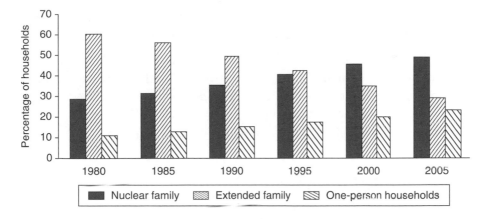

Figure 7.5  Household composition by family type, Japan 1980–2005: households with
residents aged 65+

*Source:* Japan Statistics Bureau: *Census of Japan reports 2005 and 1990*

with more couples living together into old age. However, Figure 7.5 also illustrates an increase in elderly Japanese living alone; this accounted for less than 10 per cent in 1980 rising to 23 per cent in 2005. Hence in Japan more elderly people are choosing (or being forced) to remain independent and not to live with adult children.

Despite these significant shifts in Japanese household composition in recent years, Japanese propensity to live in complex households remains distinctive from other developed societies and living arrangements in Japan interweave both traditional along with more modern practices. Budak et al. (1996) suggest four main explanations for the distinctive Japanese living arrangements:

1  *'Limping modernisation'* of Japan. One possible explanation is that Japanese cultural/social modernisation lags behind its economic development, and in time Japanese society will more closely resemble other developed nations.

2  *Independence or dependence.* An alternative explanation is that core cultural values in Japan are different from other countries. Whereas in the US and UK remaining independent for as long as possible is seen as desirable and a sign of status for the elderly, in Japan this is not the case. In contrast interpersonal dependence, where family members rely on each other, is more important.

3  *Adaptiveness of Japanese people.* This theory also stresses the mutual benefits of extended families and that living in complex families does not reflect a lack of change from historical times, but that Japanese people have adapted this form of living to suit modern life to the benefit of all involved. For example, elderly people would not be seen as a burden or dependent on their children, but as family members who make an important contribution to family life, for example looking after the children

so that both parents can work. This then is of benefit to the elderly people who are cared for when they become unable to care for themselves.

4   *Patterns of inheritance.* In some areas of Japan, people tend to pass the property rights on to one of their children while they are still alive rather than after their death. They then continue to live there with the family of the child to whom they have passed the property. This continued practice of having the eldest child continue the family line therefore makes Japan distinct from many Western countries.

## Female-headed households

As we have discussed, one of the main themes of living arrangements in historical context has been the shift towards smaller, less complex, forms. Another important trend, that can seen as part of this individualisation of living arrangements, is that households are becoming more feminised, and moreover that this trend is a global phenomenon, observed in both developing and developed societies. By the feminisation of households, we refer to the increase in female headship. We might regard the notion of household headship as somewhat anachronistic, as in the UK for example the status of the household head as a recognised figurehead with authority in the household has been replaced by more egalitarian practices that promote equality and sharing responsibilities. However, the significance of female-headed households is that these are households that are resourced by women alone. Female-headed households may be defined as households headed by a 'woman in the absence of a co-resident spouse or partner (or in some cases, another male adult such as a father or brother) (Chant, 1997: 5). Female-headed households are thus defined by the absence of an adult male, while a male-headed household is defined as one with an intact couple or other adult females, i.e. it is not defined by the absence of a woman, but taken as the norm for household formation. The basic asymmetry of these definitions implies the inferior status of female-headed households. In some societies, the assumption of a male head means that an elder son will be defined as a household head rather than an adult woman (Chant, 1997: 8). This has implications for the collection of statistics on female-headed households and for the legal status of women in this position. The gender imbalance in recognising women's status as household heads is not just a matter of statistics, but incorporates attitudes to gender relations and women's inferior position in the household. As Chant claims 'it is important to recognise that household headship is often so integrally bound with masculinity that to earn respect *as men*, males have to exert authority *as* household heads' (Chant, 1997: 8).

Given the status of female-headed households, comparison of female headship rates is not always straightforward; the number of female-headed households in a population will depend on how they are recognised both socially and legally. Some of the variation in female headship rates will reflect women's status as well as different patterns of living arrangements. For these reasons, female headship rates are not routinely collated by many national statistical agencies (see UN, 2005b). Notwithstanding these problems, female headship rates for select countries for the 1990s are given in Table 7.3, and illustrate considerable variation, with highest rates in industrialised societies (such as the Netherlands and Estonia) compared to lower rates in less developed societies (the Gambia, Mexico and Malaysia) (Chant, 1997: 70–74; Folbre, 1991). However, there are important exceptions, for example, Zimbabwe has a high rate of female headship, while in Japan the proportion of female-headed households is low, hence it is far too simplistic to associate female headship with economic modernisation. Moreover, it is important to bear in mind that higher rates of female headship will not just reflect a greater proportion of female-headed households in the population, but also different approaches to defining household heads.

Data for both more and less developed societies indicate an increase in female-headed households from the late 1970s onwards (Folbre, 1991). In the US for example, the percentage of households headed by a woman has increased from 15 per cent in 1940 to 30 per cent in 2002. This increase is associated with a rise in lone-parent families. In the US in 1950, a lone parent headed 6 per cent of families with children under 18; by 2002 this had risen to 22 per cent (Kreider, and Elliott, 2009). In developing societies, where data are available, they illustrate a general upward trend, for example in Jamaica from 28.6 per cent in 1946 to 39 per cent in 1975; Puerto Rico from 16 to 23 per cent between 1970 and 1990; and in Ghana from 22 to 29 per cent between 1960 and 1987 (Chant, 1997: 72). However, despite the evidence for growth in female headship, this does not mean that it is a 'new' characteristic of household formation. Laslett recorded

Table 7.3   Percentage of households headed by women, 1990–9

| Country | Per cent women-headed households |
| --- | --- |
| Gambia | 15.9 |
| Mexico | 16.3 |
| Malaysia | 18.5 |
| Japan | 20.0 |
| Brazil | 23.1 |
| UK | 25.3 |
| USA | 29.0 |
| Zimbabwe | 32.8 |
| Poland | 35.2 |
| Netherlands | 42.8 |
| Estonia | 54.2 |

*Source:* United Nations Human Settlements Programme (2001) *Cities in A Globalizing World – Global Report on Human Settlements* London: Earthscan. *Statistical Annexes* A-4 Ownership of Housing Units, Selected Countries

20 per cent of English households being headed by women between the sixteenth and nineteenth centuries. Historical data for Brazil also indicate a high proportion of female-headed households, as high as 45 per cent of urban households in São Paulo in 1802 (Chant, 1997: 73).

There are a number of factors that might have contributed to this global rise in female-headed households. These include economic changes, particularly increases in women's economic activity; demographic factors such as population ageing, migration; and changing patterns of family formation and dissolution (Kamerman and Kahn, 1988). Taking economic factors first, feminisation of the workforce has occurred in practically all countries in the developing and developed worlds during the latter half of the twentieth century. This suggests that more women have access to an independent source of income and are therefore more likely to be able to head their own households. However, it is important to bear in mind that women have not entered the labour market on an equal footing to men. Many are trapped in low-wage work and not necessarily in a position to support an independent household. Demographic factors are closely linked to economic modernisation, such as the rise in labour migration, which may result in women being 'left behind' to care for their families, or women may themselves be migrants. Population ageing has also had an impact, as women generally experience greater life expectancy than men; hence widowhood remains an important route of entry into households headship. In the US in 2003 for example, 14 per cent of female family heads were widows (Fields, 2003). Changes in patterns of family formation (such as the rise in never-married lone parents) and dissolution (increase in divorce) have also resulted in more households being headed by a woman. In the US, 45 per cent of female-headed non-couple households in 2003 were headed by a divorced or separated woman (Fields, 2003).

A final factor that has contributed to the growth in female heads of household is related to the overall position of women in society, which through the international women's movement has supported women's ability to take responsibility for their own households (Kamerman and Kahn, 1988). Thus it is too simplistic to conclude that female-headed households represent 'the poorest of the poor' (Henshall Momsen, 2002). Recognition of women's status by organisations such as the UN and its commitment to improving women's position, not just in terms of employment but also with regard to education, health, citizenship and legal rights, has sought to encourage governments to evaluate their position on women's roles in society. For example, changes in divorce legislation, mainly in more developed countries, have sought to recognise women's rights following marital breakdown and to facilitate women in setting up their own households (Chant, 1997). However, it remains the case that throughout both the developing and the developed world, women living on their own or with their children, are far more likely to live in poverty than women in couple households. Focusing on female headed households with children, in the US in 2003, of the 10 million households headed by a lone mother, 32 per cent had an

---

**Box 7.2    Female-headed households and
poverty: a case study of Nigeria**

Using data from the 1999 Nigerian Demographic and Health Survey, Blessing
Mberu (2007) explores the relationship between household structure and living
conditions. Female-headed households in which the woman is the only adult are
significantly poorer compared to two-parent households and male-headed single
adult households. Female-headed households with other adults do fare better
than single-female households, but are no better off than two-adult households.
However, male-headed households with other adults do benefit from the input
of additional adults. Her results illustrate the extent of gender bias in poverty,
but also the importance of considering other aspects of household structure, as
in Nigeria the number of adults in a household is a key variable in understanding
relative poverty between households.

---

average annual income below the poverty level (defined as half the median household
income, Fields, 2003).


# Non-family households

Despite the observation that most households are formed around family groups, not all
households contain a nuclear-family group. Some are based around either one person
living alone, or groups of non-related or related (e.g. siblings) individuals. However,
these types of households have not been subject to the same degree of scrutiny as family-
based households, partly because there are fewer of these kinds of households, but,
more significantly, they are generally treated as transitory arrangements, from which
members will move on to form family households. Hence, there are relatively few studies of the experiences of living in either one-person households, or house-shares (see
for example Chandler et al., 2004; Hall and Ogden, 2003; Heath and Cleaver, 2004).
Yet of all the trends in living arrangements in recent years, particularly in developed
countries, it is the increase in non-family living, particularly one-person households,
that has been most dramatic.

   One of the biggest changes to household formation in contemporary developed societies is the increase in one-person households. For example, in the UK, evidence from
the census is that the proportion of one-person households almost doubled between
1971 and 2001: in 1971, 17 per cent of households consisted of one person, by 1991
this has increased to 26 per cent. Since 1991, the increase in one-person households has
levelled off; by 2001 one-person households accounted for 31 per cent of households
(ONS, 2007). Similar trends are found in other industrialised countries, for example in

the US, 13 per cent of households consisted of one resident in 1960, by 2010 this had increased to 27 per cent (US Census Bureau, 2004).

## Characteristics of one-person households

One-person households are defined by the following characteristics:

*Age:* While it remains the case that the likelihood of living alone increases with age – for example in England and Wales, in the year 2000, 50 per cent of people aged 75 and over lived alone compared to 12 per cent of those aged 25 to 44 – the biggest proportional increase in living alone has occurred among younger age groups. In 1971, only 5 per cent of people aged 16–59 lived alone; by 2000 this had increased to 16 per cent. The breakdown of living alone by gender varies with age, with a greater proportion of older women living alone – 60 per cent of women aged 75 compared to 33 per cent of men in the same age group. However, at younger age groups, men living alone outnumber women by three to two.

*Social class and poverty:* The relationship between living alone and social class is complex. On the one hand, living alone may be regarded as a maker of economic prosperity, given the relatively higher costs in maintaining a one-person household. However, at the other end of the income scale, living alone is also associated with poverty and deprivation. For example, results from the 2001 Census for England and Wales show that:

> Single-person households are least likely to have amenities such as central heating or sole use of a bath/shower and toilet. More than one-in-eight of single-person households do not have central heating – this amounts to over 383,000 pensioners and over 430,000 non-pensioners. Over 70,000 single-person households do not have sole use of a bath/shower and toilet – 21,000 of these being pensioners. More than half of pensioners living alone have a limiting long-term illness (52.8 per cent). In pensioner-family households, 60.4 per cent contains someone with a limiting long-term illness. (ONS, 2003)

*Geography:* One-person households have a specific geography within countries. In the UK for example, we find one-person households are concentrated in more urbanised local authority districts (ONS, 2003). In 2001, for example, the City of London had the highest percentage of one-person households, with almost two-thirds of households (60.5 per cent) enumerated in the census having one resident only. In contrast in rural areas the proportion is much lower, for example in the rural southern district of Hart in Hampshire less than one-quarter (22.4 per cent) of households are one-person units.

In the US, analysis of the distribution of one-person households in the 2000 Census reveals that it is counties with the lowest proportion of married couple households that had the highest proportion of one-person households. The larger counties (those with a population greater than 100,000) of the District of Columbia; New York

County, New York; Suffolk County, Massachusetts and San Francisco County, California were all in the top ten for the proportion of one-person households and the bottom ten for married couple households (Simmons and O'Neill, 2001).

The concentration of one-person households in city centres reflects two trends, that of the wealthy affluent single householder, often in a professional occupation (such as in the City of London) who can afford to maintain their own residence (and may often have a living-apart-together partner also with his/her own home). However, as noted above in the discussion of social class, one-person households are also concentrated in poorer inner-city communities, in districts with high rates of multiple-occupancy housing.

Within cities, one-person households are not evenly dispersed, but tend to be concentrated in more central districts. Figure 7.6 gives the distribution of one-person households by ward for Liverpool in 2001. The distribution clearly shows a concentration of one-person households towards the centre of the city in inner-city wards, particularly Central, Prince's Park, Kensington, Riverside and St Michael's. Housing stock in these wards includes a high concentration of non-self contained households (e.g. bed sits or older properties divided into flats) and in general a high degree of deprivation among households living alone. These wards also house a large number of students. In contrast, wards in the suburbs such as Allerton, Childwall and West Derby have much lower rates of one-person households.

*Reasons for living alone*: Given the diversity of one-person households, the reasons for living alone are complex and reflect the extent to which individuals choose to live alone, or have no other option. More young people might choose to live alone, rather than with parents or a partner. For professional workers with the financial means to live alone, the benefits may outweigh the disadvantages, particularly if the nature of their work means that they need to be mobile. However, the experiences of affluent professionals will be very different from those of young people who are forced to leave home and end up in poor-quality housing. Another important causal factor is divorce, as marital or partnership breakdown will often, though not necessarily, mean that at least one partner lives alone immediately following separation. Finally, widowhood remains one of the main routes into living alone.

One impact of the recent changes in living arrangements is that living alone is less stigmatised. For example, the French demographer Roussel writing in the 1970s described the singleton as living on the margins of society, yet as Hall and Ogden (2003) document for single-person households in the 1990s in France and England, this description is less valid, particularly for individuals who make a positive choice to live alone.

The increase in one-person households has important impacts in terms of housing policy. For planners and politicians, an increase in one-person households means more households to be accommodated, which means more dwellings are needed. For example Table 7.4 (on page 139) gives current projections of households for England from 1991 to 2033. This shows an expectation of a larger increase in one-person households

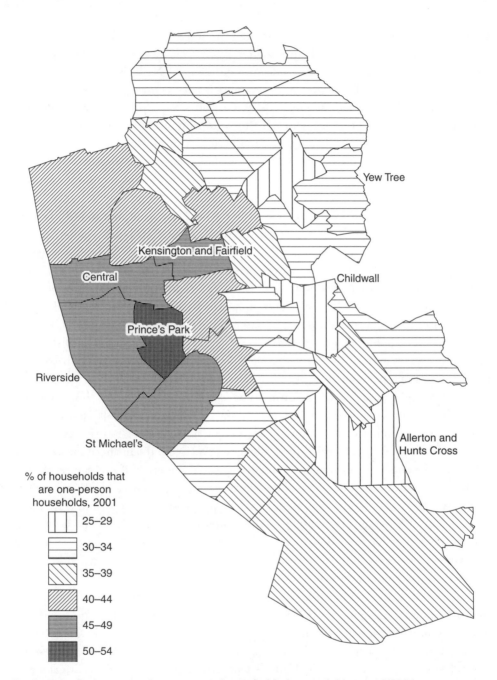

Figure 7.6    Percentage of one-person households by ward, Liverpool 2001

*Source*: *Census for England and Wales 2001*

(123 per cent) than any other household type and an overall reduction in household size. By 2033, there are projected to be more one-person households than married-couple households in England.

Table 7.4   Projected number of households by household type, England 1991–2033

| Household type | Number of households (thousands) | | | | Average annual change (1991–2033) |
|---|---|---|---|---|---|
| | 1991 | 2013 | 2023 | 2033 | |
| One family: households | 8852 | 9797 | 10327 | 10792 | 22 |
| One family: lone parent | 781 | 1585 | 2012 | 2276 | 191 |
| Other multi-person | 4479 | 3478 | 3278 | 3190 | −29 |
| One person | 5052 | 8092 | 9702 | 11279 | 123 |
| **All households** | **19166** | **22868** | **25320** | **27536** | 44 |
| Average household size | 2.45 | 2.29 | 2.22 | 2.16 | |

*Source:* Department for Communities and Local Government. Household Projections, Table 404: Household Projections 1 by Household Type and Region, England, 2001–2033, Retrieved November 2011 from http://www.communities.gov.uk/housing/housingresearch/housingstatistics/housingstatisticsby/householdestimates/livetables-households/

# Households, families and the life course

One of the limitations to approaching living arrangements through comparisons of classifications of households is that we do not get much of a sense of the processes of households and family formation. It is all too easy to regard household and families as fixed entities that remain static over time. Yet, in reality, experiences of living arrangements vary considerably, particularly over an individual's life course. Over the life course, individuals will probably spend a large proportion of time living in a nuclear family group, either as a child, a partner and/or a parent. However, they might experience living in a lone-parent family, or at other times might favour living in a non-family household, such as sharing with friends while at university, living alone during the early years of establishing a career, and living alone again following the death or departure of a spouse. Living arrangements are therefore clearly subject to considerable flux and fluidity.

One way of illustrating this is to use data on the experiences of lone-mother households over time. Analysis of household formation in Great Britain during the 1990s by Ermisch and Francesconi (2000) reveals both the frequency of partnership breakdown and re-partnering. Their analysis found that 40 per cent of mothers might expect to become a lone mother with dependent children at some point in time during their lives, yet experiences of lone parenthood are often short-lived, with one half of women remaining a lone mother for 4.6 years or less. About three-quarters of lone mothers observed went on to form a stepfamily. However, many of these are also short-lived, with one-quarter dissolved within one year. This trend towards increasing rates of both family dissolution and re-formation has important implications for children, as Ermisch and Francesconi observe:

> an increasing proportion of today's young children in Britain are likely to experience the changes, tensions and strains which life in lone-parent families and step families often entails. (2000: 235)

Hence for many parents as well as children, family and household composition cannot be taken as a fixed arrangement, but are subject to considerable change over often relatively short periods of time. We will consider the life course in more detail in the following chapter on family formation.

A final consideration is why it is important to consider variations in living arrangements. As we have suggested throughout this chapter, how we organise our lives has implications for the provision of care, as well as resourcing everyday consumption. A popular interpretation is that big extended households provide more resources, particularly for vulnerable members, while the trend towards more individualised units is symptomatic of the decline of family and support. In particular the increase in elderly living on their own has implications for how care is provided. Moreover, as outlined above, the increase in one-person households has significant impacts on the size of the housing stock required to house these smaller households. At the same time though we need to be wary of treating households as clearly demarcated and isolated boxes. The changing patterns of living arrangements for the elderly in Japan, do not necessarily signify a decline in intergenerational familial support, but a different spatial configuration of how this support is provided. We might conclude therefore that in the twenty-first century the concept of the household itself becomes anachronistic; rather, we need to recognise the fluidity of living arrangements and how the provision of care, experiences of intimacy and consumption practices extend beyond the boundaries of the household.

## Summary

- In most societies, living arrangements are based around family units within households. Household and family are therefore closely interconnected, but also distinct. We may regard the family as referring to intimate personal relationships, while the household is both a spatially and economically delimited concept. Common membership of a household is usually defined with respect to the pooling of resources and/or shared living spaces.

- In classifying family types within households, a distinction is made between simple nuclear family households and extended or complex family households. This classification has implications for provision of care for elderly and sick, as well as patterns of property ownership and inheritance.

- Variation in household size reflects both levels of fertility as well as organisational structures of households. In contemporary industrialised societies, small household size reflects both low levels of fertility, but also dominance of simple nuclear family household types.

- Despite popular beliefs about large households in past times, simple nuclear family households dominated household formation in some pre-industrial societies, particularly England. However, in more recent years household size has declined. In some societies, such as Japan, the shift from extended-family form to nuclear arrangements has been particularly pronounced in recent years.

- The proportion of households headed by women is increasing in both developed and developing societies. Causal factors associated with this include increases in women's employment, demographic factors such as labour migration and ageing, as well as changes in patterns of family formation and dissolution. The global increase in female-headed households has also been associated with the feminization of poverty.

- One of the biggest changes in household composition in modern industrialised societies is the rise in the number of people living alone. There are a number of reasons for this, including an increase in young people choosing to live alone, as well as rising divorce rates and increasing numbers of elderly people living alone. This changing pattern of living arrangements has important consequences for housing policy and provision of suitable dwelling spaces.

- Classifications of households and families may reinforce an inappropriate understanding that they are fixed entities. However, living arrangements are dynamic and individuals will experience varied and complex forms over the life course.

# Recommended reading

## Family and household characteristics

ONS (2007) *Focus on Families,* TSO: London; provides a comprehensive review of recent trends in family and household composition in the UK. The US Census Bureau website (www.census.gov/) is an invaluable source for US data. In particular the Bureau publishes an annual review of American Family Living arrangements, based on the Current Population Survey: see www.census.gov/population/www/socdemo/hh-fam.html

## Historical household and family formation

The classic reading here is Laslett, T.P.R. (1965) 'Misbeliefs about our ancestors' in *The World We Have Lost,* London: Routledge. A classic book that changed many assumptions about family life in the past). For a critique of Laslett see Michael Anderson (1995) *Approaches to the History of the Western Family, 1500–1914.* Cambridge: Cambridge University Press. Ruggles's recent review of living arrangements also argues against Laslett's argument for North European exceptionalism: Ruggles, S. (2009) 'Reconsidering the NorthWest European family system', *Population and Development Review*, 35 (2): 249–73.

## Women-headed households

For an excellent detailed account see Sylvia Chant (1997) *Women-headed Households: Diversity and Dynamics in the Developing World*. Basingstoke: Macmillan. For a more recent discussion of the relationship between poverty and female-headed households see Chant's (2004) article 'Dangerous equations? How female-headed households became the poorest of the poor: causes, consequences and cautions', *IDS Bulletin*, 35 (4): 19–26.

## One-person households

There has been very little written about one-person households, though Ray Hall and Philip Ogden have compared one-person households in France and UK, see for example Hall, R. and Ogden, P. (2003) 'The rise of living alone in Inner London: trends among the population of working age', *Environment and Planning A,* 35 (5): 871–88 and Chandler, J., Williams, M., Maconachie, M., Collett, T. and Dodgeon, B. (2009) 'Living alone: its place in household formation and change', *Sociological Research Online*, 9 (3): www.socresonline.org.uk/9/3/chandler.html.

## Life course

There has been considerably more written about the life course. For a discussion of the complexity of life course events see John F. Ermisch, Marco Francesconi (2000) 'The increasing complexity of family relationships: lifetime experience of lone motherhood and stepfamilies in Great Britain', *European Journal of Population*, 16: 235–49 and Dennis P. Hogan and Frances K. Goldsheider (2003) 'Success and challenge in demographic studies of the life course' in Jeylan T. Mortimer and Michael J. Shanahan (eds), *Handbook of the Life Course*. New York: Kluwer Academic/Plenum Publishers, pp. 681–91.

# 8

# FAMILY FORMATION

This chapter addresses the ways in which individuals make choices about events surrounding family formation. In doing so it explores how recent debates about individualisation have influenced our understanding of the process of family formation in modern industrialised societies.

Questions and issues addressed in this chapter are:

- What do we mean by individualisation, and how relevant is the concept for understanding the process of family formation in modern industrialised societies?

- What do we understand by the second demographic transition?

- How have experiences of leaving home, partnership formation and dissolution and parenting changed over the last 30 years in modern industrialised societies?

A key theme to consider, in addressing these questions, is the extent to which individuals shape their own life course events or to what extent we follow pre-determined scripts. In recent years, social scientists researching the life course have sought to unravel this tension between agency and structure, and in particular have incorporated recent ideas about individualisation.

There is no shortage of data detailing changes to family formation and fertility, particularly the 'decline' in marriage and the concomitant increase in divorce and cohabitation. For example in Great Britain in 2008:

- 52 per cent of men and 49 per cent of women aged 16 and over were married. This has declined from 74 per cent for both sexes in 1979.

- 28 per cent of men and 21 per cent of women were single.

- 3 per cent of men and 10 per cent of women were widowed.

- 6 per cent of men and 9 per cent of women were either divorced or separated.

- 11 per cent of men and 10 per cent of women were cohabiting.

- A higher per cent of younger men (aged less than 30) cohabited than older men and women.

   Source: General Lifestyle Survey, 2008

- 16 per cent of women born in England and Wales in 1923 were childless by age 45. For women born in 1973, 23 per cent of women were estimated to be childless by age 45.

   Source: ONS *Social Trends* 30

This chapter seeks to shed light on possible explanations for the emergence of new patterns of family formation and fertility and consider the complexity of contemporary family life. Before turning to consider the statistics above in more depth, we begin by exploring the concept of individualisation and what this means for the contemporary study of populations.

# Individualisation

> 'Individualisation' consists of transforming the human 'identity' from a 'given' into a 'task' – and charging the actors with the responsibility for performing that task and for the consequences of their performance. (Bauman, 2002: xv)

Individualisation is a key theme in understanding everyday lives and experiences in modern industrialised societies. Writers such as Anthony Giddens, Zygmunt Bauman, Ulrich Beck and Elisabeth Beck-Gernsheim address the ways in which modern life is becoming more 'individualised' and the consequences of this for both the individuals concerned and society at large. The focus of individualisation is, as the term suggests, the individual actor and the ways in which individuals are actively engaged in constructing their own lives. It is theorised that individuals in modern industrialised societies are becoming less burdened by the restrictions that influenced their parents' or grandparents' lives. In particular, social structures such as class and gender are less restrictive, and individuals have more opportunities and choices (or at least the perception of more opportunities) to pursue their own life-goals. The emphasis on opportunity and choice suggests that individual actors are less constrained by the traditions of the past; and expectations that we organise our lives around certain ways of doing things, because that is how they have always been done, are becoming less relevant.

   A key theme that writers have used to explore the distinctive nature of individualisation is that of the self-negotiated or do-it-yourself biography, which places the individual very much in the driving seat of their own life trajectories. Hence when Bauman writes of human identity becoming a 'task', rather than something that is 'given', the emphasis is on how individuals perform this task, rather than acquiesce with what is given. Another important dimension of the do-it-yourself biography is that it is a reflexive experience; individual actors are continually re-assessing their own lives, and re-writing their own scripts in order to respond to changing situations. Trajectories can, and will, change over time. The emphasis on the self also implies, as Bauman argues, that at the end of the day, the individual can be held to account for their own actions, successes and failures. The logical extension of individualisation is that people can no longer blame others (e.g. parents) or structures (e.g. social relations based on class and gender) for their own life experiences and outcomes.

So what does the experience of individualisation mean for the study of population? Population events, from birth through leaving home, partnership formation (and dissolution), and parenthood are key stages in individual biographies, and the ways in which individuals experience these events are becoming increasingly characterised by the process of individualisation. We might usefully use the concept of the 'do-it-yourself' biography to understand how individuals make decisions about demographic events. A key theme to explore is the ways in which individuals are free to make their decisions regarding key life events, such as getting married or having children, or whether the legacy of embedded social relations, such as class, ethnicity and gender, continues to have an influence on individual behaviours. For example, gender relations are a key factor in the processes of family formation. As Beck and Beck Gernsheim argue, one of the main tensions arising out of the do-it-yourself biography is the way in which men and women reconcile their own individual needs and expectations with those of their immediate family:

> The tension in family life today is the fact that equality of men and women cannot be created in an institutional family structure which presupposes and enforces their inequality. (Beck and Beck-Gernsheim, 2002: xxii)

Family lives continue to be influenced by gendered social relations, which influence patterns of employment both within and outside the home. We therefore need to consider the extent to which women are increasingly 'free' to negotiate their own biographies, particularly relating to combining work and motherhood. How are women's opportunities to construct their own biographies limited by expectations that they will perform gendered tasks within the home, such as taking on the responsibility for housework, particularly childcare, while combining this with their own personal employment aspirations outside the home? We also should consider how men's expectations of fatherhood are shaped by changing gender relations.

This chapter asks how relevant is individualisation as an organising theoretical principle that can be applied to contemporary experiences of family formation. While this chapter focuses on family formation, we should be aware that the theme of individualisation is relevant for all demographic events. The ideal of individual choice and responsibility underscores key debates in migration and health inequalities as well as family formation.

# Second demographic transition

The concept of individualisation has been developed by social scientists to explore the process of social change. In the study of populations, there are a number of trends in the characteristics of demographic events, which we may loosely group together as being part of the same overall process of change. Demographers such as Lesthaeghe (1995) and Van de Kaa (1987) use the concept of the second demographic transition to describe the changes that are characteristic of populations that have passed through the final stages of the first demographic transition (see Chapter 2). The second demographic transition links key life history events with trends in employment patterns and gender relations in the home and the workplace. By linking these events and treating them as part of the same overall transition, apparently distinctive changes can be considered as part of the same overall processes. The main changes that make up the process of the second demographic transition are given in Box 8.1.

---

### Box 8.1    Characteristics of the second demographic transition

- Decline in fertility
- Decline of male bread-winner model (i.e. man goes out to work, woman stays at home)
- Decline in marriage
- Increase in childless or one-child couples
- Increase in women's employment
- Increase in cohabitation
- Increase in dual-career families
- Increase in one-parent families

---

The second demographic transition does not just focus on describing demographic events, but how these events are experienced and the extent to which determinants of individual behaviour during the second demographic transition are distinctive from those of the first transition. Lesthaeghe identifies a change from altruistic behaviour that characterised the first transition to individualised behaviour of the second transition. In particular, at the time of the first demographic transition couples were motivated to limit fertility to reduce the risk of infant mortality, in other words a change in behaviour occurred to protect the welfare of children. This contrasts with behaviour during the second demographic transition that is far more individualised. At the time of the second demographic transition, individuals are motivated to ensure their own personal well-being, and rights to self-realisation are prioritised, particularly for women.

We can see, therefore, that both demographers and sociologists have identified how the focus on individual actors and the choices that they make is key to understanding current patterns of family formation in modern industrialised societies. We shall now explore these patterns in more detail, by focusing on three key stages of individual life histories: leaving home, partnership formation and dissolution and parenthood.

# Leaving home

Leaving home is an important transition in young people's lives, yet one that, until recently, has been mostly ignored by population researchers. This is partly due to the assumption that leaving home occurs as a consequence of other transitions, such as getting married, starting work or going to university, as opposed to a transition of interest in its own right. However, in recent years in many modern industrialised societies, leaving home has become more disconnected from other events, particularly partnership formation. Moreover, leaving home is, more often than not, a process rather than a one-off event, with many young people returning to the parental home. These changes in the experiences of leaving home have stimulated more interest in understanding the choices and constraints that influence young people's behaviour.

## Trends and characteristics of leaving home transitions

One of the most interesting aspects of leaving home in modern industrial societies is the variation both over time and place. Taking temporal trends first, in the UK patterns of leaving home have varied considerably during the twentieth century (Jones, 1995). In the 1950s and 1960s, for example, most young people left at relatively young ages (in their early twenties) and, except for the minority who went to university, they left home to get married. This pattern of leaving home to get married is commonly thought of as the 'traditional' pattern of leaving home in the UK, yet data from the eighteenth and late nineteenth century reveal a more complex picture. Prior to the twentieth century, for many young people leaving home was a protracted process, with many leaving at young ages (leaving home in the early teens was not uncommon) to work as apprentices in new industries or as domestic servants; very few left to get married. Settling down in a marital household was usually delayed until young people had the resources to establish themselves independently, often in their mid- to late-twenties (Wall, 1978; Pooley and Turnbull, 2004). Hence, we need to be cautious in taking the experiences of young people in the 1960s as an exemplar of 'traditional' practices, when the youth labour market was buoyant and the housing market relatively cheap.

The pattern of leaving home characteristic of the post-Second World War era began to change in the mid-1970s, as the youth labour market began to contract and more young people, especially women, were encouraged to continue their studies in higher education (Jones, 1995). At this time, we see a trend back towards more protracted transitions with more young people leaving home prior to marriage and the forma-tion of independent family households. This trend has continued into the twenty-first century, with fewer young people leaving home to get married, or even live with a partner.

Trends in the proportion of young people living with parents by age and gender in the US are given in Figure 8.1. Since the 1960s, the trend towards higher rates of co-residence between young people and parents is apparent, though this does appear to have been reversed in the 1990s, before increasing again in the early years of the twenty-first century. This recent increase in living at home has been more discernable for young women under 25. The increase in the number of American young people living at home in the latter half of the twentieth century has received considerable attention from the media and parents themselves, as well as academic researchers

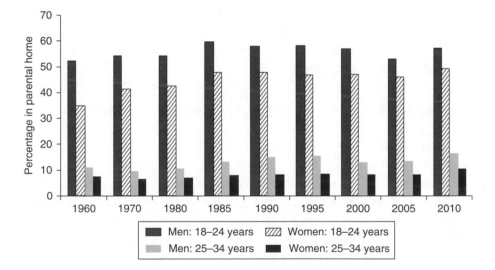

Figure 8.1    Percentage of young American adults living in the parental home by age and gender, 1960–2010

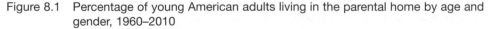

*Source:* US Census Bureau (2004) Table AD-1. Young Adults Living At Home: 1960 to Present Retrieved December 2010 from www.census.gov/population/socdemo/hh-fam/tabAD-1.pdf

(Goldscheider and Goldscheider, 1999) and a number of terms, some of them quite derogatory, have emerged to describe young people's and parents' experiences of delayed leaving home. Young returnees are described as 'boomerang kids', or 'incompletely launched adults'. Parents may also lament the fact that the 'empty nest', which they were looking forward to after their children left home, is being denied to them.

It is not just change over time that is a significant characteristic of leaving home, as geographical variation is also important. There is considerable variation in the living arrangements of young people throughout industrialised societies (Billari et al., 2001; Iacavou, 2002 and Holdsworth, 2000). Data for the average age of leaving home for men and women in 2007, as collated by the European Union, is illustrated in Figure 8.2. It can be seen that leaving home is geographically specific, with younger ages in the North, while in the South and East of Europe average ages of leaving home are much older. It is interesting to note that there appear to be two groupings of countries, with Estonia and Lithuania demarking the division between countries with younger ages of leaving home compared to societies with older average ages. In Mediterranean countries, it is not unusual for young people to live with their parents in their late twenties and early thirties. Conversely in Northern Europe, particularly Scandinavian countries, the likelihood of a young person in their late twenties living in the parental home is relatively small. As in North America, in all European countries women leave home at earlier ages than men.

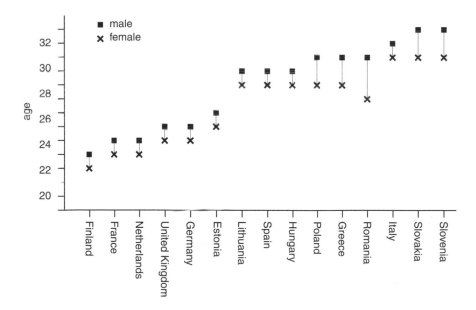

Figure 8.2   Average ages of leaving the parental home by gender, select European countries 2007

*Source:* Eurostat (2009) *Youth in Europe a Statistical Portrait*

## Leaving home and individualisation

What do these trends of leaving home across time and space tell us about the experience of leaving home, and, in particular, how it relates to our understanding of individualisation? On the one hand, the recent trends in the reasons for leaving documented for both the US and UK would appear to endorse the process of individualisation, in particular that young people are rejecting 'traditional' leaving home trajectories (i.e. leaving home to get married), and instead embarking on a variety of trajectories, which include leaving home prior to partnership formation and, increasingly, returning to the parental home. Yet, as the historical data for England illustrate, we need to be wary of labelling transitions for the post-Second World War era as 'traditional', given that the experiences of today's generations of young people have more in common with young people in the late eighteenth and nineteenth centuries, than they do with the experiences of generations leaving home in the 1950s and 1960s. However, we could argue that what is distinctive about contemporary transitions is the way in which young people have some choice about leaving home, for example whether to leave home to go to university, which is an integral element of the 'do-it-yourself' biography. However, again, we need to be cautious in over emphasising the amount of choice that young people have today, given the nature of the youth labour and housing market (Ford et al., 2002).

The situation in Southern Europe would also appear to suggest that 'traditional' patterns of leaving home continue to dominate young people's experiences. Again, we find that among the factors influencing the distinctive pattern of leaving home in Spain and Italy, lack of alternative options is an important theme. Iacovou's (2002) comparison of young people's transitions throughout Western Europe identifies a number of structural factors that might account for delayed leaving in countries such as Italy and Spain. These include: the lack of employment options for young people (youth unemployment in Spain in 2012 exceeded 50 per cent); the shortage of cheap housing (80 per cent of homes are owner-occupied in Spain, and cheap rented accommodation is in short supply) and the structure of the welfare state, which provides very little assistance for young people, on the assumption that their needs will be met by other family members, usually parents. The latter is particularly important, as the role of the family also influences relationships between parents and adult children (Holdsworth, 2004). This raises another possibility, that older ages of leaving home are not just a result of lack of options for young people to leave home at younger ages, but also reflect a real choice: that young people are happy living with their parents, for the most part, and can see no reason for leaving home until they are ready to get married and establish an independent household. Hence, the process of individualisation in Southern Europe is somewhat complicated. At the very least the concept of choice appears to be constrained by structural factors and as well as cultural practices of family life.

---

### Box 8.2    Biographies of young people leaving home

*Carlota*: My name is Carlota and I'm 28 years old and I live in Bilbao in Northern Spain, where I work as a lawyer. I've just left home to get married and I live with my husband Jon, who I've known since I was 18. My parents live in a remote part of rural Galicia, in North-West Spain, and I moved away from Galicia when I was 12, to come and live with my sister and brother-in-law in Bilbao. I moved away in order to continue my studies, as I always wanted to be a lawyer, but the schools where my parents lived were not that good.  I lived with my sister and her family for approximately 16 years, most of time I shared a room with my niece, Rosa. My sister and her husband, Esteban, have been like a mother and father to me. When I was a teenager I used to argue with Esteban about what time I had to be home by, he was stricter than my dad! Despite this I never thought about leaving home, even when I met Jon. My sister and Esteban weren't that keen about me spending time with Jon at first, I remember the first time I went on holiday with him, there was a lot of discussion between my sister, Esteban and my parents, but in the end they allowed me to go, I was 23 years old at that time! Despite this, I've been really happy living with my sister and her family, though I love living with Jon, I miss my old home, I still have a key and go back whenever I want to!

*Amber*: My name is Amber, I'm 22 years old and I live on my own in a rented flat in Liverpool. I'm not working at the moment, as I'm doing a course in order to go to University next year. I first left home when I was 17, after a huge row with my Mum. It was really dramatic; I remember throwing all my things into plastic bags, my sister was screaming at me and my Mum was really upset. I had to apply for benefits when I left, but my Mum helped me out as well, even though she wanted me to return home. Since then I've lived in all sorts of places, I even went to work as an au pair in Sweden, and spent some time down in London working in a hotel, but that didn't work out; so I came back to Liverpool. I've had some really scary experiences since I left home, including one time when this guy was stalking me, so I had to leave my flat, and go back home. My Mum has always let me go back home when I've needed to, and despite the rows, she has always helped me out. Now I'm settled in my little flat, I really want to stay, but I'm worried about the cost of going to University next year, as I will lose most of my benefits. I don't know if I will be able to keep this flat, I'll just have to work at the same time as studying, as I really don't want to go back home.

## Partnership formation

One of the most distinguishing characteristics of the second demographic transition is what is happening to partnership formation in modern industrialised societies. As the data outlined at the beginning of this chapter illustrate, in 2008 less than half of adult women in Great Britain were married. So if men and women are not getting married, what is happening? There are a number of factors that influence the decline in marriage:

- delay in age at first marriage; more young people are remaining single for longer;

- increase in cohabitation, either as a trial or an alternative to marriage;

- increase in divorce, though this is offset by increase in remarriage.

## Timing of marriage

In the 1950s and 1960s, most young people in the UK got married in their early twenties yet nowadays many young people may delay getting married until their mid-thirties, with an increasing number of people deciding not to get married at all. Data on median age at first marriage for women given in Table 8.1 demonstrate this trend for other European countries and the US (conventionally demographers have focused on the timing of key family formation events for women, as the ages that women get married/have children influence levels of fertility. This convention is used in this chapter). The extent of delayed marriage varies with women in Western Europe marrying at older ages than their counterparts in the US. We also find larger increases in age at first marriage in Northern Europe (e.g. Sweden), compared to the South (e.g. Italy), in part because women were getting married in countries such as Italy in the 1960s at relatively older ages. Recent increases in median age at marriage (from 1995 to 2000) are also worth noting. However, the trend of delayed marriage is not uniform. As the data for Russia illustrate, in parts of Eastern Europe age at first marriage has remained relatively young and, at least in Russia, has declined since the mid-1960s.

## Cohabitation

Part of the decline in marriage in modern industrialised societies is due to the increase in cohabitation. This in part reflects a generation effect, as cohabitation is now more acceptable, but also an age effect as young people choose to cohabit prior to marriage (Kiernan, 2004). It is unclear whether people are choosing to cohabit as a trial of marriage or as an alternative to marriage. It is likely that both are the case, and that the increase in cohabitation contributes to the delay in marriage, as much as an absolute decline in the number of couples getting married.

As with leaving home and the timing of marriage, there are important geographical differences in cohabitation. Figure 8.3 illustrates geographical variation in the extent to which women have embraced cohabitation as a substitute for marriage, comparing European countries. Women who have left home in Northern Europe, particularly in Scandinavia, are far more likely to be cohabiting than in Southern Europe, where

Table 8.1   Median age at marriage for women, select countries 1965–2000

|          | 1965 | 1975 | 1985 | 1995 | 2000 |
|----------|------|------|------|------|------|
| France   | 22.7 | 22.5 | 24.2 | 26.9 | 28.0 |
| Germany  | 22.9 | 22.3 | 24.1 | 26.4 | 27.0 |
| Italy    | 24.2 | 23.7 | 24.5 | 26.6 | 27.4 |
| Russia   | 24.0 | 22.7 | 22.2 | 22.0 | N/A  |
| Sweden   | 23.5 | 24.8 | 27.2 | 28.7 | 30.2 |
| UK       | 22.6 | 22.5 | 23.9 | 26.4 | 27.5 |
| US       | 20.6 | 21.1 | 23.3 | 24.5 | 25.1 |

*Source: Council of Europe (2004) Demographic Year Book.* Strasbourg: Council of Europe Publishing.
US Bureau of the Census (2006) Table MS-2. Estimated median age at first marriage, by sex: 1890 to the present. Retrieved December 2010 from www.census.gov/population/socdemo/hh-fam/ms2.pdf Accessed December 2010.

Data are based on US current population survey

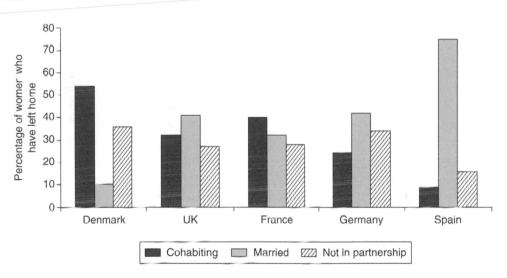

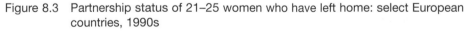

Figure 8.3   Partnership status of 21–25 women who have left home: select European
            countries, 1990s
*Source:* Iacovou and Berthoud (2001)

cohabitation remains relatively unusual, In fact, the situation is Southern Europe is very distinctive, as marriage remains the preferred option.

# Divorce

Divorce rates vary considerably between countries. Divorce rates for European countries are given in Figure 8.4. Divorce rates are highest in countries in Northern

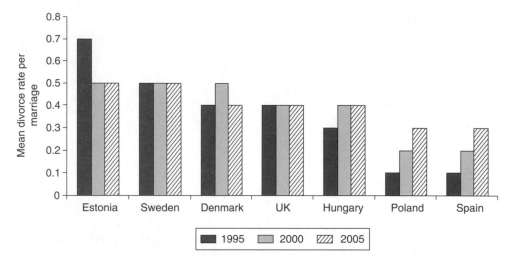

Figure 8.4   Divorce rates for select European countries, 1995, 2000 and 2005

*Source:* Eurostat Data service, European Commission

Europe and parts of Eastern Europe and are lower in Southern Europe and 'Catholic' Eastern Europe (Poland). The distribution of the frequency of divorce is, therefore, similar to that for cohabitation, though it is noteworthy that divorce rates have increased noticeably in Spain and Poland between 1995 and 2005.

However, high divorce rates do not indicate a decline in marriage. Data from the US (Box 8.3) illustrates that while more people are getting divorced, they are not 'falling out of love' with getting married.

Unlike leaving home or getting married, the assumption of choice as applied to divorce is somewhat different, as we might expect that it is not a decision that people make willingly. Despite the increasing fashion for pre-nuptial agreements, when most couples get married they do not envisage marriage ending in divorce. In order to provide both couples and policy makers with more awareness of the factors that lead to divorce, there has been a considerable amount of research into the causes of marital breakdown. Identifiable risk factors include:

• Age at marriage: couples marrying at younger ages are at greater risk.

• Family history of divorce: divorce begets divorce.

• Financial difficulties and unemployment increase the likelihood of marital breakdown.

• Reliance on state benefits: poorer couples are at greater risk.

• Disability increases the chances of divorce.

   (Kiernan and Muellar, 1999)

---

### Box 8.3  Divorce and marriage in the US

- Most adults have married only once
- Most people who have ever divorced are currently married
- First marriages which end in divorce last 7 to 8 years on average
- Half of those who remarry after divorce from a first marriage do so within about 3 years

Source: Kreider and Fields (2001)

---

Other more contested risk factors include women's employment, presence of children, gender roles and cohabitation prior to marriage. To date, research on these factors illustrates contradictory results. For example, earlier research in the UK shows that presence of children reduced the risk of divorce, while more recent analysis suggests a reversal of this relationship. Likewise, attempts to identify the relationship between cohabitation prior to marriage and risk of divorce reveal considerable variation in different cultural contexts (Kulu and Boyle, 2010).

Research on the impact of female employment, and consequently higher wages, has been influenced by economic models for understanding family behaviour, influenced by the writings of the Nobel-prize winning economist Gary Becker. Becker (1981) explored how a narrowing of the wage and education gap between the sexes may increase marital instability. There is empirical research that confirms this hypothesis (e.g. Chan and Halpin, 2003), though other commentators argue that the model is over-simplistic and ignores non-economic factors (Mason, 1997).

## Divorce and individualisation

Most of the identified causal factors of increased risk of getting divorced illustrate the continual importance of structural factors, such as labour market characteristics, which continue to have an impact on marital breakdown. Yet writers such as Giddens, Beck and Beck-Gernsheim and Bauman argue that rising divorce rates illustrate the important shifts that have been taking place in people's lives and intimate relationships. What is important here is the extent to which people are willing to commit to long-term relationships. According to Beck and Beck-Gernsheim the 'traditional' certainties of family life have declined, and while people still yearn for close intimate relationships, at the same time they have more freedom to do what they want, in other words they have

more choice. Giddens also argues that the nature of partnerships is changing. He identifies a new form of intimacy based on two related concepts: confluent love, which stands
in opposition to romantic love and monogamous, heterosexual marriage; and the pure
relationship which is a 'social relationship which is entered into for its own sake, for what
can be derived by each person from a sustained association with another; and which is
continued only in so far as it is thought by both parties to deliver enough satisfactions for
each individual to stay within it' (Giddens, 1992: 58).

The new form of intimacy brought about by these two concepts is therefore a direct
challenge to 'permanent relationships' associated with romantic love and the belief that
we are going to meet a compatible partner, fall in love and stay together for the rest of
our lives. Giddens argues that these changes are coming about because in modern
societies individuals, and especially women, understand themselves differently. We are
becoming more reflexive in how we think about ourselves and our relationships, and, as
such, are conscious agents of change. Though Giddens recognises that the heterosexual
marriage (as the basis of the family) remains hampered by gendered power relations, he
argues that the fundamental challenge to partnership and family formation derives from
the emergence of new forms of relationships, and that we should not necessarily view
rising divorce rates as a pernicious aspect of late modernity.

These accounts of how individualisation works for individuals are increasingly being
challenged by empirical studies of marriage and divorce. In particular, researchers
investigating the experiences of divorced parents and step-parents describe how this
view of moving from one pure relationship to another is far from straightforward when
children are involved (Smart and Neale, 1999; Smart, 2007). In these situations, parents
have to re-negotiate their relationships with their ex-partners, family and children,
rather than simply move on to a new relationship.

## Divorce and impact on children

One of the main concerns that both parents and policy makers have to face when dealing with divorce is the impact that it might have on children. There is no conclusive
evidence as to how children react to parental divorce, but numerous studies, mainly
carried out in the US and UK, have identified a number of outcomes including:

> lower education performance;
> more disruptive behaviour;
> more likely to leave home early;
> children of divorcees are more likely themselves to get divorced.
> (Elliott and Richards, 1991; Ní Bhrolcháin et al., 2000 and Cherlin, 1999)

However, there is considerable debate among social scientists as to the precise reasons
why children may experience these outcomes. We can potentially explain the relationship
between marital dissolution and outcomes for children in four ways:

1  The outcomes do indeed reflect children's experience of divorce and the fact that children have to come to terms with, and live through, the breakdown of their parents' marriage. However, this should not necessarily be viewed in a negative light. For example, the fact that divorce begets divorce may be taken as evidence that children of divorcees find it harder to maintain relationships as adults, or, alternatively, that having lived through it once, they are more likely to be open to the possibility of divorce, and see it as a solution to marital problems. In reality, the ways that children deal with divorce are very complex, and we cannot predict how and why children react to the actuality of parental separation.

2  The relationship between childhood behavioural problems and divorce is the reverse of that which is commonly assumed, i.e. it is not parental separation that leads to behavioural problems, but these behavioural problems predate divorce and these are a contributing factor to marital breakdown.

3  The outcomes reflect avoidable aspects of divorce, such as having to change schools if parents have to move (hence the impact on educational performance), or moving house.

4  The outcomes have little to do with divorce *per se*, but reflect the experiences of children living in an disharmonious home prior to divorce. It is unlikely that their parents' relationship was cordial before the marital split, and the experience of living with conflict can be distressing for children. (Ní Bhrolcháin et al., 2000; Ní Bhrolcháin, 2001).

The need to mitigate the potential (though not necessarily proven) deleterious outcomes of divorce on children has influenced divorce reform in many modern industrial societies in recent years. For example, some states in the US have introduced no-fault divorce on the basis that if divorce is unpleasant, particularly for any children involved, then it is better to make the experience of divorce as non-confrontational as possible. In the UK, the emphasis of recent legal reforms has been on the potential to save 'save-able' marriages, and reducing conflict where marital breakdown is irretrievable. Parents are encouraged to come to agreements about joint-parenting without necessarily getting the courts involved, and both parents have a legal responsibility for their children after divorce (Smart and Neale, 1999).

Since the 1990s, a considerable amount of research has been carried out on the impact of divorce on family members, mainly in North America and the UK. Divorce is clearly an emotive issue. Some commentators regard divorce as a significant causal factor of contemporary social problems, while others equate divorce with opportunities for freedom and choice. As the American demographer Paul Amato concludes, from what we know about divorce neither argument can be fully substantiated, rather we have to recognise the complexity of divorce and its outcomes:

The increase in marital instability has not brought society to the brink of chaos, but neither has it led to a golden age of freedom and self-actualisation. Divorce benefits some individuals, leads others to experience temporary decrements in well-being that improve over time, and forces others on a downward cycle from which they might never fully recover. (Amato, 2000: 1282)

# Explaining nuptiality differences across Europe

There are important differences in marriage, cohabitation and divorce across Europe, though these are not new. In a landmark paper first published in 1965, John Hajnal described a distinctive difference in historical patterns of marriage and household formation, distinguishing Western and Eastern Europe. Hajnal's line ran from St Petersburg in Russia to Trieste in North-east Italy: to the east of the line, marriage was early and almost universal, while to the west, marriage was later and many people never married. Hajnal associated this difference in marriage timing with distinctive household formation: in the East complex households were more common, with young couples living in their parents' home rather than establishing their own independent household; in the West nuclear family households were more common.

These cultural traditions continue to shape contemporary marriage patterns, though are mediated by other factors. Kalmijn's (2007) analysis of marriage behaviour in Europe concludes that gender roles and religion are the most influential explanatory variables. Increases in women's employment outside the home are associated with low marriage rates, high levels of cohabitation and high divorce rates. Marriage is more widespread and divorce is less common in more religious countries. Economic indicators are less important: unemployment reduces marriage, but does not affect the timing of marriage. Finally, education is the least important variable, though higher education (attending university) is associated with higher rates of both marriage and divorce. This analysis also illustrates the importance of cultural practices, particularly when considering the timing of marriage, as Kalmijn concludes:

> The degree of historical continuity and the regional differences are especially marked for marriage timing, which may point to the persistence of marriage customs over time. (Kalmijn, 2007: 259)

# Parenting

One of the biggest choices that we make in our adult lives concerns children: whether to have children, when to have children, how many children to have. Availability of contraception in modern industrialised societies means that most, though not all, individuals are in the position to be able to make these kinds of decisions about having children. In fact, we find that the fact that having a child is a 'choice' nowadays reveals an interesting paradox. As fewer couples choose to have children, and the proportion of childless women is increasing in most modern industrialised societies, among women who cannot, for whatever reason, 'choose' to have children, more are turning to medical treatments to enable them to conceive. The rapid expansion of reproductive technologies and the large amount of money that is spent on research and the development of new techniques, as well as by couples on these treatments, illustrates how 'fertility' really is big business. The number of books, newspaper and magazines articles, extolling the virtues of either being 'childfree' or of being a 'parent' also evidences the dilemma of the choice of parenthood. The zeitgeist of modern industrialised societies is that fertility decisions have to be justified.

## Trends in fertility

There are three important trends in fertility in modern industrialised societies:

*Childlessness:* An increasing number of women are choosing not to have children. The level of childlessness depends on both the number of involuntary childless women (those who are unable to have children) and voluntary childless women (who choose to remain childless). As the level of involuntary childlessness will remain fairly stable over time (though expansion of fertility treatments in recent years will lead to a decline in involuntary childlessness), increases in childlessness are due to more women choosing not to have children. As data on childlessness given in Table 8.2 illustrate, most industrialised societies have witnessed an increase in childlessness in recent years, though the level of childlessness varies between different countries. In the UK for example, over 20 per cent of women born in 1960 remain childless, and this compares with around 7 per cent of women in France.

*Births outside marriage:* One of the important trends has been the decline in marriage as the conventional setting for having children. As more couples are cohabiting, we find that trends in births outside marriage are increasing. Trends in the percentage of births

Table 8.2    Percentage of women childless by birth cohort (year of women's birth)

|             | 1940 | 1945 | 1950 | 1955 | 1960 |
|-------------|------|------|------|------|------|
| UK          | 11   | 10   | 14   | 17   | 21   |
| France      | 8    | 7    | 7    | 8    | 8    |
| Finland     | 14   | 14   | 16   | 18   | 18   |
| Netherlands | 12   | 12   | 15   | 17   | 18   |
| Spain       | 8    | 6    | 10   | 10   | 10   |
| Ireland     | 5    | 3    | 9    | 14   | 16   |

*Source:* Pearce (1999)

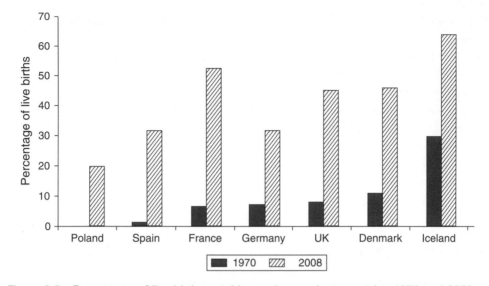

Figure 8.5    Percentage of live births outside marriage, select countries 1970 and 2008

*Source:* Data taken from OECD (2011) Family Database, OECD, Paris © OECD. Retrieved December 2011 from www.oecd.org/social/family/database

born outside marriage for select European countries are illustrated in Figure 8.5, comparing 1970 and 2008. It can be seen that the increase in the proportion of births outside marriage has been a consistent trend in most modern industrialised countries. The greatest proportion of births to unmarried parents are found, not surprisingly, in those countries with the highest cohabitation rates, i.e. Scandinavian countries, and the lowest rates in Southern Europe, where cohabitation is less common. However, even in Southern Europe there has been a noticeable trend towards more births outside marriage in recent years, particularly in Spain, where the proportion is similar to that for Germany.

*Women's mean age at childbirth:* Among women who do have children, more are choosing to have their children later. Again, as we can see from Table 8.3 this is a common trend in most industrialised societies. While in 1970 the average age of women at childbirth varied from 24.6 in the US to 29.6 in Spain, by 2005 this had increased to 27.4 in the US and 30.9 in Spain. Hence while the overall trend is in the same direction, the start and end points are quite different. Women in Spain had their children, on average, at older ages in 1970 compared to women in the US in 2005.

## Trends in fertility in global context

The focus of this discussion is the more developed world first. However, we should not overlook similar trends and debates in a wider global context. In particular, the advantages

Table 8.3    Mean age of women at first childbirth

|         | 1970 | 1980 | 1990 | 2000 | 2005 |
|---------|------|------|------|------|------|
| Germany | 26.6 | 26.4 | 27.6 | 28.7 | 29.6 |
| Spain   | 29.6 | 28.2 | 28.9 | 30.7 | 30.9 |
| France  | 27.2 | 26.8 | 28.3 | 29.3 | 29.7 |
| Sweden  | 27.0 | 27.6 | 28.6 | 29.9 | 30.5 |
| UK      | 26.3 | 26.9 | 27.7 | 28.5 | 29.2 |
| US      | 24.6 | 25.0 | 26.4 | 27.2 | 27.4 |

*Source:* Eurostat Data Service, European Commission; Mathews and Hamilton (2002) and US National Centre for Health Statistics website

of older parenthood are recognised in very different contexts. In the developing world, as in the global north, rising enrolment in post-primary education; improvements in medical practice and knowledge; the impact of economic modernisation and associated urbanisation; and the expansion of civic society are all associated with changes in the expectations of parenthood and the prerequisites for young people to be considered culturally 'mature' to be parents. However, data for the global distribution of fertility by age reveals a very varied pattern.

Data on the contribution to total fertility by women in two age groups, distinguishing between mothers aged less than 30 and those 30 and over, illustrate that the trend towards older ages of motherhood is more pronounced in Europe and North America than in less developed world regions, see Table 8.4. This is because in the earlier period,

Table 8.4    Percentage of total fertility contributed by women aged 15–29 and 30–39, 1990–95 and 1995–2000 by world region

|                             | Percentage of total fertility contributed by women aged | | | |
|-----------------------------|-------|-------|-------|-------|
|                             | 15–29 | 30–49 | 15–29 | 30–49 |
| World region                | 1990–95 | | 1995–2000 | |
| Africa                      | 56    | 44    | 57    | 43    |
| Asia                        | 65    | 35    | 65    | 35    |
| Europe                      | 72    | 28    | 65    | 35    |
| Latin America and Caribbean | 66    | 34    | 66    | 34    |
| Northern America            | 69    | 31    | 65    | 35    |
| Australia/New Zealand       | 63    | 37    | 57    | 43    |

*Source:* United Nations Department of Economic and Social Affairs Population Division (2004) *World Population Monitoring 2002 Reproductive Rights And Reproductive Health.* New York: United Nations

younger women accounted for a larger proportion of births, particularly in Europe, compared to other world regions. The end result of this trend has been a more even distribution of fertility by age across the different global regions, as the distribution in Europe and North America by the end of the twentieth century has converged towards the existing pattern in Asia and Latin America. Africa remains distinctive though, with a more even distribution of fertility across the life course. This reflects cultural practices of childbearing that emphasise spacing of births throughout a woman's reproductive lifespan, in contrast to the more bounded stopping behaviour adopted in European societies.

## Women's employment and family formation

The complexities of fertility choices do not relate just to *whether* and *when* to have children but also to *how* individuals organise their lives to respond to their children's needs. For women in particular, fertility decisions are closely interwoven with choices about jobs and careers. The kinds of considerations that influence fertility decisions include: when is the best time to take a break from employment to have children; whether to return to work when children are young; whether to work full or part-time. These issues increasingly apply both to men and women, though we shall consider the relationship between women's unemployment and fertility.

Data on trends in women's employment illustrate how, despite the fact that there has been an increase in women's employment in most countries, there are still tensions between women's involvement in paid work and their roles as mothers and partners within the home. One potential outcome of the tensions that women experience in combining work with family life, is that they have to choose between the two. The difficulty of combining work and motherhood is a very topical issue in modern industrialised societies, and has had profound impacts not just on women, but also on their families and society in general. In particular, as the increase in women's employment has occurred at the same time as declining fertility, there is a consensus that there is a negative relationship between fertility and women's employment (Bernhardt, 1993). Yet the precise casual mechanism of this relationship is less clear, and there is much discussion about the extent to which employment reduces fertility, but also how having children impacts on women's employment options (Joshi, 2002).

One of the problems in defining the relationship between fertility and women's employment is that, if we look at trends both over time and place, the relationship is not consistent. Though most countries in Europe and North America have experienced a decline in fertility at the same time that women's employment has increased, data for the current levels of fertility and employment reveal a complex relationship (Brewster and Rindfuss, 2000). For example, if we focus on Europe, a curious pattern emerges: it is the countries with the *lowest* rates of female employment that currently have the *lowest* rates of fertility (Esping-Andersen, 1999; Kohler et al., 2006; Frejka and Sobotka,

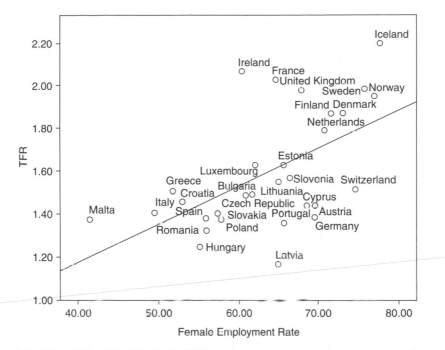

Figure 8.6   Plot of Total Fertility Rate (TFR) against percentage of women in employment aged 20–64, 30 European countries 2010

*Source:* Eurostat Data service, European Commission (no date) Main Tables

2008). This relationship is illustrated in Figure 8.6 which gives a plot of total fertility rate against the female employment rate for 2010. The best-fit line for the relationship between fertility and employment is also plotted on the graph. European countries are loosely divided into three groups. In the top right-half of the graph are countries char-acterised by higher rates of fertility (TFR above 1.7) and female employment (female employment rate above 60 per cent). These are all Northern European, mostly Scandinavian countries. The outlier in this group is Iceland with a TFR of 2.2 and female employment rate of 78 per cent. The second group of countries includes Southern and Eastern European countries, all of which have a TFR below 1.6 and a female employment rate below 60 per cent. Countries in this group include Spain and Poland, both with low TFRs (1.38) and a low female employment rate (56 and 58 per cent, respectively). The final group is a more disparate group though mainly consists of Central and Eastern European countries with low fertility (TFR less than 1.65) and higher rate of female employment. An outlier in this group is Switzerland which has a high rate of female employment (75 per cent) and low rate of fertility (TFR of 1.52).

In trying to unravel the complexities of the relationship between women's employ-ment and fertility, it is useful to consider the problem from both perspectives; that is, to explore how fertility impacts on women's employment, before turning to consider the influence of employment on fertility.

## Fertility impacts on employment

Let us start by considering the ways in which women's fertility might impact on employment options. The extent to which having children impacts on employment will depend on options to combine the two. The contradictions between motherhood and employment will depend on age of the child, and will be most acute for mothers of young children. Data from the US for women with children under one illustrates the immediate impact of fertility on women's employment. As illustrated in Figure 8.7, among women who had their first child between 1961 and 1965, less than 20 per cent had returned to work a year after giving birth; by 1991–4 this has increased to 60 per cent. Hence, while it is reasonable to expect fertility to have an immediate impact on women's employment opportunities, in countries such as the US this impact has diminished over time.

The extent to which fertility reduces women's employment options will depend on a number of factors, in particular, access to affordable childcare and 'family-friendly' jobs which allow women and men to balance the demands of work and family.

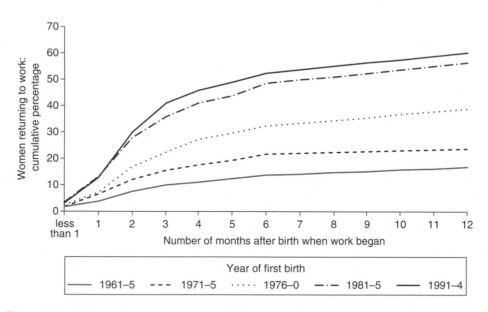

Figure 8.7   Women who returned to work after first birth by year of first birth: US

*Source:* Smith, Downs and O'Connell (2001). Data reproduced by kind permission from US Census Bureau, graph is author's own interpretation

## Family-friendly employment

One of the most important factors that impacts on women's opportunities to combine motherhood and employment is the type of work that they, and also their partners, are engaged in. In recent years there has been considerable emphasis on encouraging

family-friendly workplaces, which recognise that working parents often have conflict-
ing demands and need to be flexible in their working arrangements to respond to this,
for example, when a child is ill or during school holidays. According to the OECD,
family-friendly policies are those:

> that increase resources of households with dependent children; foster child
> development; reduce barriers to having children and combining work and family
> commitments; and, promote gender equity in employment opportunities.
> (OECD, no date)

However, within this broad definition there are a number of different strategies. At one
extreme, polices may be introduced which facilitate women leaving the labour force,
through generous maternity pay and leave and provision of support for women looking
after children. At the other end of the spectrum are polices that provide women with
childcare to enable them to continue working. Associated with this are more passive
measures which endorse the need for flexible workplace arrangements to facilitate balancing work with the needs of their children. These include enabling time off to look
after children, options of changing hours worked to fit in around school hours and
parental leave for both mothers and fathers (Castels, 2003).

One of the most important strands of family-friendly employment is the provision of
childcare. The type and amount of childcare provided for young children varies considerably in different countries (Chesnais, 1996). The most common form of childcare in
many developed countries, and the only form of childcare in developing countries, is
that provided by other family members, such as partner, grandparents or older siblings.
Even in more industrialised economies access to formal childcare is limited. For example,
data on childcare arrangements given in Table 8.5 for the US illustrate the importance of
family-provided care as well as organised care, with half of all women with pre-school
children relying on family members to look after their children.

Table 8.5   Primary childcare arrangements used by employed mothers of pre-school
            children, 1999: USA

| Type of arrangement | Per cent distribution |
| --- | --- |
| *Family and relatives* | *50.3* |
| Mother while working | 3.1 |
| Father | 18.5 |
| Grandparent | 20.8 |
| Sibling and other relative | 8.0 |
| *Organized facility* | *22.1* |
| *Other non-relative care* | *27.6* |
| Total children under 5 years | 10,587 |

*Source:* Laughlin (2010)

For women who do not rely on relatives, less than half use organised facilities such as day care centres, with the remainder using other forms of childcare, mainly childminders.

Availability of state-funded childcare varies considerably geographically. Figure 8.8 compares data collated by the OECD to indicate variation in the percentage of GDP that different countries spent on pre-school education and childcare in 2003. It is interesting to note that Iceland and Switzerland are at either ends of this graph, as both countries have high female employment rates, but very different levels of fertility, with Iceland recording the highest fertility level in Europe.

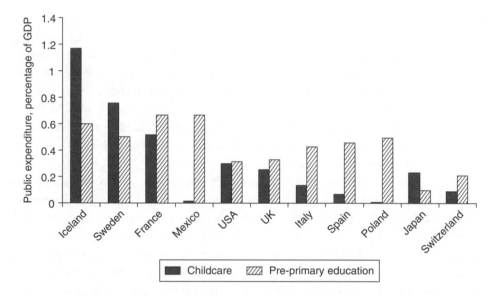

Figure 8.8    Public expenditure on childcare and early education services, percentage of of GDP, 2003

*Source:* Data taken from OECD (2007) Social Expenditure Database 1980–2003. © OECD. Retrieved December 2009 from www.oecd.org/els/social/expenditure

## Employment impacts on fertility

While the impact that having children has on women's employment is not consistent across time or place, there is at least evidence of how fertility shapes employment opportunities. However, if we turn to look at the reverse relationship then the effect of employment on fertility is a far more contested issue (Bernhardt, 1993). In attempting to understand how fertility decisions are influenced by employment, we need to disentangle plans and behaviour surrounding employment and fertility, and, explore how couples respond to anticipated events and conditions in the future. If a childless working woman expects to have problems combining motherhood and employment then we would expect this to have a subduing effect on her fertility plans. Bernhardt's review of the research on the impact of employment on fertility plans concludes that the available evidence as to how employment impacts on fertility is inconclusive. More-over, as we have seen above, conditions for combining motherhood and employment vary considerably over time and place, and it is reasonable to assume that the introduction of family-friendly employment practices and the extension of childcare provision in many industrialised countries, particularly in Europe, will have altered the way in which fertility plans are mediated by employment options.

## Whither individualisation?

While this chapter has emphasised the importance of choice, it remains the case that not all individuals in modern industrialised societies are free to entirely choose their own life paths. An important critique of commentaries on individualisation is that they play down the significance of constraints on individual decisions. We need therefore to consider the extent to which the do-it-yourself biography is applicable for everyone. Some of the possible constraints that have been suggested in this chapter include: gender, education and locality, while other ones to consider are social class, ethnicity and migration status. Cultural practices are also proving more resilient to social change and in some aspects of family formation, particularly leaving home and age at marriage, we can see that continuity in behaviours over time remains important (Kujisten, 1996). Finally, as Carol Smart (2007) has discussed, individuals are not leading more atomised lives. While they might experience more flux in their personal lives, individuals remain tied to family members (both current and former) and friends. For example, separated parents are expected to maintain relationships with their children and ex-partners in order to provide supportive parenting for their children. Hence, family separation does not necessarily mean the severing of personal ties but the creation of more complex linkages.

More recent trends in family formation and fertility practices in Europe also challenge the universality of the second demographic transition. In particular, there are important differences in the relationship between trends in female employment, fertility, leaving home and partnership formation in different parts of Europe. To assume that family formation practices in Spain will resemble those in Scandinavian countries is too simplistic. While men and women are responding to changing patterns of employment and expectations of self-realisation throughout modern industrial countries, they are responding in different ways which incorporate distinctive cultural practices and economic circumstances in their localities.

## Summary

- The concept of individualisation, which focuses on the ways in which individuals construct their own biographies (the do-it-yourself biography), is relevant for understanding family formation in modern industrialised societies. The most systematic way in which demographers have engaged with the concept of individualisation is via the idea of the second demographic transition, which relates trends in family formation with changes in patterns of employment and gender relations.

- Leaving home transitions are becoming increasingly diverse and protracted in most modern industrialised societies, yet we also find considerable diversity in different countries; in particular, leaving home patterns in Southern Europe are distinctive, as most young people do not leave until their late twenties/early thirties.

- Partnership formation is characterised by a decline in marriage, which is in part offset by an increase in cohabitation (though marriage still remains the norm). Cohabitation may be seen as a trial or an alternative to marriage.

- The increase in divorce is one of the key characteristics of family life in developed societies in the late twentieth and early twenty-first centuries. Commentators such as Giddens predict that we need to come to terms with new forms of intimacy, which preclude 'live-together-forever' relationships. The reasons for this increase in divorce are complex, and reflect both legal reforms that have facilitated divorce as well as attitudinal shifts. However, structural factors, such as unemployment, are also important risk factors for divorce.

- One specific area of concern is the impact that divorce has on children. Research has identified a number of childhood outcomes associated with marital dissolution though establishing causality is contentious.

- Decisions about parenthood may be regarded as the choice of modern industrialised societies. Choices are made about: if to have children, when, how many, and how parents (particularly women) combine their parental duties with employment.

- The relationship between women's employment and fertility is complex and has changed in recent years. Country-specific contexts are important in explaining the relationship between fertility and employment.

- Having a child has an impact on women's employment, particularly in the months immediately following giving birth. However, in recent years more women are returning to work earlier after having a child illustrating that the impact of fertility on employment is not fixed but determined by a number of factors. These include access to affordable childcare, legislation promoting family-friendly employment and maternity leave. There is considerable geographical variation in the availability of formal support for mothers, particularly working mothers. In many countries, informal provision of childcare is more important than formal provision.

- The impact that employment has on fertility is difficult to establish. Couples may choose to delay or forgo having children if they are concerned about the difficulties of combining work and family life, yet it might be quite difficult for women to make a judgement as to what the impact of having a child will be.

# Recommended reading

## Individualisation

The key 'texts' of each of the 'key' writers on individualisation are:

Giddens, A. (1992) *The Transformation of Intimacy: Sexuality, Love and Eroticism in Modern Societies*. Cambridge: Polity Press.

Beck, U. and Beck-Gernsheim, E. (2002) *Individualisation: Institutionalized Individualism and its Social and Political Consequences*. London: Sage.

Bauman, Z. (2001) *The Individualised Society*. Cambridge: Polity Press.

For an overview of how these impact on family formation see:

E. Beck-Gernsheim (2002) *Reinventing the Family*. Cambridge: Polity Press.

For a critique of individualisation see:

Mason, J. (2004) 'Personal narratives, relational selves: residential histories in the living and telling', *The Sociological Review*, 52 (2): 162–79; and Carol Smart (2007) *Personal Life*. Cambridge: Polity Press.

## The second demographic transition

The key texts here are: Lesthaeghe, R. (1995) 'The second demographic transition in western countries: an interpretation', in K. Oppenheim Mason and A. Jensen (eds). *Gender and Family Change in Industrialized Countries*. Oxford: Clarendon Press.

Van de Kaa, D.J. (1987) 'The second demographic transition', *Population Bulletin*, 42 (1): Washington, DC: The Population Reference Bureau. Lesthaeghe has recently published an article (2010) 'The unfolding story of the second demographic transition', in *Population and Development Review*, 36 (2): 211–51.

More recently Buzar et al. have used this concept to explore recent changes in household formation see Buzar, S., Ogden P.E. and Hall, R. (2005) 'Households matter: the quiet demography of urban transformation', *Progress in Human Geography*, 29 (4): 413–36.

## Leaving home

For an overview of European trends see:

Iacovou, M. (2002) 'Regional differences in the transition to adulthood', *Annals of the American Academy of Political and Social Science*, 580 (1): 40–69.

Billari, F.C., Philipov, D. and Baizán Muñoz, P. (2001) 'Leaving home in Europe: the experience of cohorts born around 1960', *International Journal of Population Geography*, 7 (5): 311–38.

For a more qualitative approach see:

Holdsworth, C. (2004) 'Family support and the transition out of parental home in Britain, Spain and Norway', *Sociology*, 38 (5): 909–26.

## Marriage, cohabitation and divorce

Kath Kiernan's (2004) paper on 'Unmarried cohabitation and parenthood in Britain and Europe', *Law and Policy*, 26 (1): 33–55 provides a thorough overview of recent trends and her chapter with Muellar (1999) in Susan McRae (ed.) *Changing Britain: Families and Households in the 1990s*. Oxford: Oxford University Press gives an overview of divorce characteristics in the UK.

Matthijis Kalmijn's paper on nuptiality is Kalmijn, M. (2007) 'Explaining cross-national differences in marriage, cohabitation, and divorce in Europe, 1990–2000', *Population Studies*, 61 (3): 243–63.

The impact of divorce on children is summarised in the following:

Ní Brolcháin, M., Chappell, R., Diamond, I. and Jameson, C. (2000) 'Parental divorce and outcomes for children', *European Sociological Review*, 16 (1): 67–92.
Cherlin, A. (1999) 'Going to extremes: family structure, children's wellbeing and social science', *Demography*, 36 (4): 421–8.

## Fertility

Eva Bernhardt's paper provides a good review of the relationship between fertility and employment, Bernhardt, E. (1993) 'Fertility and employment', *European Sociological Review*, 9: 25–42.
On mothers' employment see: Joshi, H. (2002) 'Production, reproduction, and education: women, children, and work in a British perspective', *Population and Development Review*, 28: 445–74. Long but informative.
Gotha Esping-Andersen discusses the relationship between fertility and employment in his 1999 book *Social Foundations of Post-industrial Economies*. Oxford: Oxford University Press.
This relationship is also discussed in Francis Castels's 2003 paper: 'The world turned upside down: below replacement fertility, changing preferences and family-friendly public policy in 21 OECD countries', *Journal of European Social Policy*, 13 (3): 209–27.

# 9

# HEALTH INEQUALITIES

The chance of an individual living a long and healthy life varies according to a number of factors including their country of residence, their socio-economic circumstances and the particular neighbourhood in which they live. This chapter explores health inequalities between places and population groups addressing the following questions:

- Why is it important to measure population health?
- What is health?
- How can we measure the health of a population?
- What is the extent of health inequalities between countries?
- What is the extent of health inequalities between population groups and areas in the UK?
- Is the health of an area the result of the type of people who live there or the type of area they live in?
- Can the level of wealth inequality in a country influence the health of the population?

## Why is it important to measure population health?

Health is important: the desire to live a long and healthy life is an almost universal aspiration. However, the chances of people meeting this aspiration are far from equal. The statistics below illustrate the extent of inequalities both internationally and within the UK:

- In 2009, a baby born in Swaziland could expect to live until the age of 49 whilst a baby born in Switzerland could expect to live until the age of 82 (UN Population Division).

- In 2009, 68 per cent of all HIV infections were found in Sub-Saharan Africa (UN Global report on AIDS – 2010).

- In 2009, 24 per cent of adults in Zimbabwe were HIV positive (UN Global report on AIDS – 2010).

- The total excess deaths in the most disadvantaged half of the population in the UK that result from the higher levels of mortality (compared to the most advantaged half of the population) is equivalent to a major air crash each day (Benzeval et al., 1995).

- A boy born in the relatively deprived Calton area of Glasgow has a life expectancy of just 54 – 28 years less than one born a few miles away in Lenzie (WHO, 2008).

- In the UK, a child of an unskilled father is twice as likely to die before the age of 15 than a child with a professional father (Benzeval et al., 1995).

From a moral perspective, the existence of health inequalities across areas and social groups, such as those listed above, is a concern for those who believe that everybody, regardless of the social circumstances they are born into, should have an equal chance of living a long and healthy life. Attitudinal surveys suggest that most people perceive the extent of inequality between social groups in the UK to be a problem. For example, the British Social Attitudes Survey reports that 78 per cent of people in Britain felt the income gap between rich and poor people was too large in 2009 with this proportion fluctuating around 80 per cent between 1986 and 2009.

From a practical perspective, the study of population health is important as it informs the ways in which national governments and organisations such as the UN and the World Health Organisation tackle health inequalities. In some areas, research on population health gives cautious grounds for optimism. For example, there has been a reduction in the prevalence of HIV in some of the worst affected African countries in recent years. However, it also clear that improvements in the health of people in the poorest countries and of the poor in all countries are not happening quickly enough. It is also true that many of the health inequalities we measure today are remarkably persistent. Research shows that the least healthy areas of Inner London have remained the same for the last 100 years (Dorling, 2000). Therefore, an additional role for population health researchers is to find evidence to support the implementation of new policies to address health inequalities and to monitor the extent of health inequalities in the future.

## What is health?

Health can be defined in a number of ways. A large body of research demonstrates that health means different things to different people and that concepts change over time.

Those in the richest countries might feel that their sore throat makes them 'ill' but this sentiment is unlikely to have been shared by their equivalents living 100 years ago or those living in the poorest countries today.

Blaxter (1990) examines research on lay perceptions of health finding four dimensions: absence of illness, ability to function, fitness and the idea of health as a reserve which is diminished by self-neglect and accumulated through healthy behaviour (e.g. exercise). The extent to which each health dimension is drawn upon in an individual's conceptualisation of health varies between people and social groups. In this sense, we can regard health as a socially constructed phenomenon – our ideas of what it means to be healthy depend upon our expectations which are influenced by a range of factors such as our neighbours, the media, the government and our access to healthcare. The contested nature of what health is leads to a number of definitions being developed and used in health research.

Since 1948 the UN have defined health as:

> Health is a state of complete physical, mental and social well-being and not merely the absence of disease or infirmity. (UN, 2010)

This definition gives a somewhat idealised vision of health and accordingly its utility is somewhat limited as under this definition most people would consider themselves as unhealthy for most of the time (Gatrell and Elliott, 2009)! Alternatively, health could be defined according to the ability of an individual to function effectively and participate within society. A weakness of this approach is that many people have illnesses that do not interfere with everyday activities or have developed strategies to participate fully in everyday life and so might not be picked up by this definition. For example, Blaxter (1990) cites evidence from interview studies on health where respondents define themselves as healthy whilst at the same time suffering severe disease or incapacity.

From a purely medical perspective, the presence or absence of disease as diagnosed by a medical professional is used to define whether a person is healthy or not. This definition of health is compromised in that a person may feel ill but might not have a disease or biological abnormality that is detectable by a health professional. For example, Senior and Viveash (1998) report the findings of a study on doctors' and patients' perspectives on the diagnosis of repetitive strain injury (RSI), a condition that covers a number of neck, shoulder, hand, wrist, forearm and elbow disorders. They note the opposition to the diagnosis of RSI among some doctors because there seems to be no identifiable physical cause. However, patients' perspectives on RSI are both valid and in some cases have changed doctors' opinions on the seriousness of the condition.

The complexity of health as a concept can be further illustrated through debates around the notion of mental illness and consideration of the reasons for its increasing prominence in affluent countries such as the UK. Senior and Viveash (1998) provide a useful critique of the debates around the concept of mental illness that is summarised here. One of the main challenges is that definitions of mental illness inevitably rely on a value judgement as to what determines abnormal behaviour and which of the abnormal behaviours should be classified as a mental illness. For example, homosexuality was once classified as a mental disorder, but as a result of pressure from gay and lesbian rights groups it is now properly seen as a choice or orientation rather than an illness. Judgements on mental illness are usually made by the medical profession or powerful groups such as politicians and so tend to reflect the experiences of the more privileged sectors of society and their views of normality.

In spite of the challenges associated with defining and measuring mental health, mental illness is widely recognised as a major social and economic problem in many countries, including the UK. Mental health problems account for more than 12 per cent of the total NHS budget (McCrone et al., 2008) and depression is the most common single cause for long-term sickness (after back problems) (Dorling, 2010). According to the 2007 Adult Psychiatric Morbidity Survey (APMS), 15 per cent of adults had a common mental disorder which includes different types of depression and anxiety. Of those suffering from a common mental disorder, half had a condition that was of a severity that required treatment. Psychosis, which covers the most serious mental conditions with the capacity to produce disturbances in thinking and perception severe enough to distort perception of reality, affected 0.4 per cent of survey respondents.

The rise in prominence of mental health problems is likely to be due to a number of factors. For example, the last 20 years has seen increasing awareness and social acceptance of mental health problems. This has led to more people being willing to identify themselves as being mentally ill than in the past where there was greater stigma associated with such conditions. At the same time, there have been increases in the numbers of identifiable mental illnesses and disorders meaning that doctors now spot conditions that would not have been recognised as illnesses in the past. Under this theory it is not the underlying rate of mental illness which is increasing, rather that the trends reflect improvements in our abilities as a society to recognise such conditions.

Other researchers have suggested there may also be real increases in the prevalence of mental illness that are attributable to aspects of modern living that are harmful to mental health. For example, Dorling (2010) presents a theory of how increasing economic inequalities in the UK have led to increases in depression and anxiety as a consequence of individuals' attempts to enjoy the lifestyle enjoyed by the most affluent. Longer working hours, adjustments to the changing societal roles of men and women and increases in debt and job insecurity are all thought to play a part in increasing levels of mental illness.

The key point of this discussion on concepts of health is that there is no gold standard definition of what health means. When examining statistics on population health, it is important to consider the definition of health that is being used. Results would almost certainly be different were a different definition of health to be examined. The next section examines various approaches to measuring health.

## How can we measure population health?

One of the most common ways to measure population health is through records of deaths. Mortality statistics are particularly useful because information on deaths including cause of death is usually (but not always) recorded by governments. In England and Wales, it is a legal requirement to register any death and so information on mortality is very accurate compared with other statistics. The Office for National Statistics releases a set of data known as the Vital Statistics each year which includes data on deaths distinguishing cause of death, age and sex.

Although mortality has traditionally been used to measure demand for health and welfare services, it is increasingly recognised that, in developed countries (and many in Asia and Latin America), measures of morbidity (illness) over the lifetime are a valuable alternative source for assessment of well-being and the need for services (McCracken and Phillips, 2005). For example, such measures are particularly useful in situations where two people die at similar ages but have very different health experiences throughout their lives. As premature mortality becomes less common in developed countries, measures of self-assessed health are increasingly important for analysis of inequalities in health (Mitchell, 2005).

Many censuses and surveys now include questions in which respondents provide an assessment of their own health. In the UK, the census of population has been carried out since 1801, during which time sporadic questions on self-assessed health and disability have been asked (Charlton, 2000). In 2001, the census included three questions on health/disability. The self-assessed general health question allowed respondents to classify their health as good, fair or poor. The census economic activity question included a category of permanently sick or disabled. Finally, the limiting long-term illness question recorded whether an illness, health problem or disability limited an individual in their daily activities (see Box 9.1).

The estimated rates of poor health will clearly vary according to the particular question that is used. It is important to be aware that a number of methodological factors will also influence the prevalence of morbidity (illness) that is measured within a census or survey. For example, the population-sampled affects estimates of ill health because certain groups, such as the elderly or those in certain types of institutions, are at greater risk of suffering from a disability (Dale and Marsh, 1993). Some surveys allow proxy

---

## Box 9.1    A selection of self-assessed health questions

**Census (2011) – limiting long-term illness question (England)**
Are your day-to-day activities limited because of a health problem or disability
which has lasted, or is expected to last, at least 12 months?
Yes, limited a lot
Yes, limited a little
No

**Census (2001) – limiting long-term illness question**
Do you have any long-standing illness, health problem or disability which limits
your daily activities or the work that you can do? Include problems which are
due to old age.
Yes
No

**Census (2001) – General health question**
Over the last 12 months would you say your health has on the whole been:
Good
Fair
Bad

**Health Survey for England – General health question**
How is your health in general? Would you say it was:
Very good
Good
Fair
Bad
Very bad

---

answers, while others allow them for certain age groups or where a person is too
disabled to participate themselves. Surveys that allow proxy answers tend to have lower
estimates of disability than those that do not due to the proxy respondents missing low
level disabilities in others (Bajekal et al., 2003). The context of a survey can affect estimates of disability prevalence. The Health Survey for England (HSE) contains a wide
range of questions designed to measure the health of participants. It also includes a
general question on limiting long-term illness, an identical form of which was found in
the General Household Survey (GHS, a UK government survey that ran from 1971 to
2006). Estimates of limiting long-term illness are higher in the HSE (26 per cent among
adults in 2001) compared with the GHS (23 per cent among adults in 2001) and it is
thought that the health focus of the HSE is responsible for this difference – people are

more likely to define themselves as ill if they have answered a large number of health questions in the run up to the limiting long-term illness question.

Self-assessments of health are not only influenced by an individual's physical condition but also by their expectations and the comparisons that they make with the population around them. Both expectations and comparisons are culturally determined and this has led to criticism of the objectivity of self-assessed health/disability measures and in particular their use for the comparison of the prevalence of illness over time and between places (Mitchell, 2005).

Several studies express this criticism by comparing self-assessed measures of disability/ health with mortality between places and over time. Mitchell (2005) compares LLTI and life expectancy across all districts in Britain to show that for a given life expectancy Scots are less likely to report an LLTI than the Welsh. O'Reilly et al. (2005), in a similar way, demonstrate that the relationship between LLTI and mortality varies across UK regions.

In spite of these criticisms, it is important to note that a large body of research supports the validity of self-reports of disability and health (Idler and Benyamini, 1997). These measures have been shown to be significantly and independently associated with specific health problems, changes in functional status, recovery from periods of ill health, mortality, GP consultations, in- and out-patient visits (Dale and Marsh, 1993; Bowling, 2005). Self-assessed measures have been shown to be valid across the age range and for different ethnic groups in Britain (Chandola and Jenkinson, 2000; Manor et al., 2001; Bowling 2005). In the UK, it is argued that the census measures of self-reported health and disability give the only nationally consistent indication of health-service needs (Dale and Marsh, 1993).

In many countries governments sponsor the collection of a number of nationally rep resentative surveys that focus on various topics including health. Unlike the census, surveys contain rich information on the characteristics and health of individuals. In the UK, the Health Survey for England is collected on an annual basis and is used to provide reliable information about various aspects of people's health and to monitor selected health targets. The survey includes a nurse visit and medical assessments of health such as blood pressure. Health surveys are also collected in Scotland, Wales and Northern Ireland. More details on survey data on health can be found in Higgins et al. (2010). The main disadvantage of survey data is that information on the area of residence is suppressed, or restricted to large areas within a country to prevent individuals being identified. This limits the utility of survey data for analysis of health inequalities between subnational areas.

Administrative datasets on ill health include GP records (Charlton, 2000), hospital episode statistics (Price, 2000), disability registers and benefits data (Norman and Bambra, 2007). Administrative data are a rich source of information and are often available with fine geographical data offering a useful indicator of relative health for small areas

(Norman and Bambra, 2007). The key disadvantage of administrative data is that they only enumerate the population using a particular service and do not count those who may be suffering from poor health but are not on a register, claiming benefits, visiting a doctor or a hospital (Bajekal, 2000). Changes to administrative procedures limit the usefulness of such sources for time series analysis. Caution should be exercised over the claims that administrative statistics provide a more 'objective' measure of health compared to self-assessed measures. For example, whilst eligibility for a disability benefit requires a medical examination for eligibility, cultural factors such as the willingness of people to visit the doctor and to apply for such benefits complicate matters (Norman and Bambra, 2007).

Chapter 4 discussed the importance of population age structure and when reporting any health statistic for a population it is important to use a measure that controls for the age structure of the population. A more elderly population is likely to have a higher proportion of deaths, GP visits or self-assessments of poor health purely because older people are most prone to health problems as a consequence of the ageing process. Chapter 5 introduced and defined various age standardised measures of mortality including standardised mortality ratios (SMRs) and life expectancies. We can also calculate equivalent age standardised measures for illness – standardised illness ratios (SIRs) and healthy life expectancies.

The standardised illness ratio is calculated in the same way as the standardised mortality ratio except that age-specific rates of illness rather than mortality are used. A selected illness schedule is applied to the age structure of the populations to be compared giving the *expected* number of people with an illness in each of the populations. The ratio of the *observed* number of people with an illness and the *expected* number of people with an illness gives the standardised illness ratio (as for the SMR it is conventional to multiply by 100):

$$\text{SIR} = \frac{\text{Observed population with an illness (in population of interest)}}{\text{Expected population with an illness (in population of interest)}} * 100$$

Table 9.1   Crude % of population with a limiting long-term illness (LLTI) and number of people with an LLTI in England, Derbyshire Dales and Bury (2001)

| Country | % with LLTI in 2001 | Number with LLTI in 2001 |
| --- | --- | --- |
| England | 17 | 8369174 |
| Derbyshire Dales | 22 | 11703 |
| Bury | 18 | 32374 |

*Source:* ONS Census 2001 Table ST16

Table 9.2   Calculations for SIR, Derbyshire Dales and Bury

| Age Group | Population 2001 Derbyshire Dales Column 1 | Population 2001 Bury Column 2 | Age-specific LLTI rate England 2001 Column 3 | Expected population with an LLTI Derbyshire Dales Col 1* Col 3 Column 4 | Expected population with an LLTI Bury Col 2* Col 3 Column5 |
|---|---|---|---|---|---|
| 0–4 | 5405 | 11108 | 0.03 | 165 | 338 |
| 5–9 | 6261 | 12216 | 0.05 | 290 | 565 |
| 10–14 | 6188 | 12992 | 0.05 | 301 | 632 |
| 15–19 | 5557 | 10972 | 0.05 | 282 | 558 |
| 20–24 | 4811 | 8801 | 0.06 | 284 | 520 |
| 25–29 | 6017 | 11363 | 0.07 | 410 | 774 |
| 30–34 | 7307 | 13660 | 0.08 | 593 | 1109 |
| 35–39 | 7779 | 14956 | 0.10 | 771 | 1482 |
| 40–44 | 7159 | 12723 | 0.12 | 880 | 1565 |
| 45–49 | 6422 | 11663 | 0.16 | 1003 | 1822 |
| 50–54 | 7194 | 13145 | 0.20 | 1425 | 2603 |
| 55–59 | 6040 | 10437 | 0.26 | 1585 | 2739 |
| 60–64 | 4973 | 8935 | 0.34 | 1679 | 3016 |
| 65–69 | 4508 | 7707 | 0.38 | 1708 | 2920 |
| 70–74 | 4339 | 6487 | 0.44 | 1894 | 2832 |
| 75–79 | 3832 | 5296 | 0.53 | 2016 | 2787 |
| 80–84 | 2396 | 3254 | 0.60 | 1448 | 1967 |
| 85+ | 1523 | 2381 | 0.71 | 1086 | 1698 |
| Total | 97711 | 178096 | | 17821 | 29927 |

*Source:* ONS Census 2001 Table ST16

An example of the calculation of SIRs in Bury and Derbyshire Dales is given below. Here the illness schedule for England is applied to the local population structure in each district in order to calculate expected populations with an illness.

The calculation of SIRs for Bury and Derbyshire Dales are given below (see Table 9.1 for observed totals and Table 9.2 for the calculations of the expected totals). Whilst Table 9.1 shows that the crude percentage of LLTI is higher in Derbyshire Dales than in Bury, after we account for age we see that the SIR is higher in Bury than Derbyshire Dales. This is because Derbyshire Dales has a more elderly age structure than Bury which inflates the crude percentage of LLTI in this district. After we control for local age structure, levels of illness are higher in Bury. As Bury has an SIR above 100, this means it has higher levels of illness than England, after controlling for local age structure, with the converse being true for Derbyshire Dales.

SIIR (Bury) = (Observed pop with LLTI/Expected population with an LLTI)*100

SIR (Bury) = (32,374/29,927)*100

SIR (Bury) = 108.1

SIIR (Derbyshire) = (Observed pop with LLTI/Expected population with an
                   LLTI)*100

SIR (Derbyshire) = (11,703/17,821)*100

SIR (Derbyshire) = 65.7

---

### Box 9.2    Sullivan's Method

Let:

$HLE_x$ = Healthy life expectancy at age x

$P_x$ = illness rate at age x (x=1,2,3........84)

$L_x$ = person years lived at age x

$S_x$ = Number of survivors at age x

Then:

$$HLE_x = \frac{\sum_{x}^{n}((1 - p_x)L_x)}{S_x}$$

---

Healthy life expectancy is an extension to the life expectancy statistic giving the number of years a person can expect to live in a healthy state. Conventional life tables can be extended to include information on healthy life expectancies using a technique known as Sullivan's method. Sullivan's method requires only age-specific rates of morbidity (illness) which are used to partition life expectancies into years with and without illness (see Box 9.2). Jagger et al. (2006) provide an excellent practical guide for the calculation of estimates of healthy life expectancy.

A weakness of Sullivan's method is that it does not take into account movements between states of illness and health. Multistate models address this weakness using information on health transitions to model healthy life expectancies more comprehensively. However, data on the health of individuals over time that are needed to inform transition probabilities are scarce and so multistate models are not commonly used. Research suggests that the Sullivan method can, generally, be recommended for its simplicity, relative accuracy and ease of interpretation (Jagger et al., 2006).

## Inequalities in health between countries

Levels of mortality are the most reliable way to assess and compare the health experiences of the populations living in different countries. Self-assessed measures of health are less commonly used because health expectations vary so widely between countries (particularly between high and low income countries). In Chapter 5, statistics on life expectancy and rates of infant mortality in a selection of countries were displayed demonstrating the current inequalities in mortality between countries. Rates of childhood mortality (under five) are particularly useful because they are the main cause of low life expectancy in a country with high levels of mortality. The fourth millennium development goal (MDG) set by the UN focuses on infant mortality pledging to 'Reduce by two thirds between 1990 and 2015, the under five mortality rate'. Here we examine how the inequalities in mortality outlined in Chapter 5 have developed over time.

Life expectancy has increased dramatically over the last 100 years. In 1900, world life expectancy was around 30 years; however, by 2008 this had more than doubled so that the average person could expect to live to 70 years. This is a staggering achievement and it is not confined to the richest countries – more than 85 per cent of the world's population can now expect to live to at least the age of 60, twice the world life expectancy in 1900.

Figure 9.1 illustrates the changes in life expectancy in a selection of countries from 1950 to 2008. It is possible to divide countries into three groups on the basis of changes to their life expectancies between 1950 and 2008. In the richest countries, such as the US and the UK, life expectancies have improved steadily from around 70 in 1950 to 80 in 2008. With the exception of Afghanistan, all the countries with the lowest life expectancies in 2008 are in Sub-Saharan Africa. Some of these countries (e.g. Zimbabwe) actually witnessed a decline in life expectancies between 1950 and 2008, although it is important to note that the average life expectancy across all Sub-Saharan African countries increased from 40 years to 50 years during this period. The remaining countries whilst diverse in many respects have in common a dramatic increase in life expectancy leaving them closer to the most developed countries in 2008 than in 1950. China and Mexico each had a life expectancy of 50 in 1950 which had risen to around 65 by 2008. The Epidemiological Transition Model discussed in Chapter 2 provides an explanation for the changes in mortality described above linking economic development to changing causes of mortality from those that predominantly affect the young to those that predominantly affect the old.

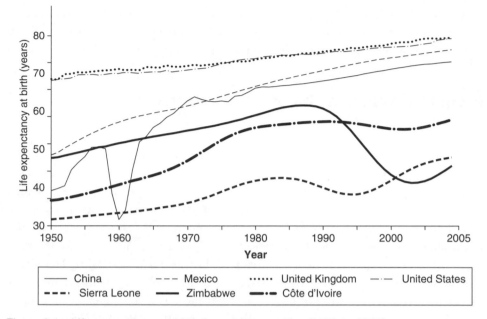

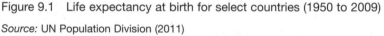

Figure 9.1   Life expectancy at birth for select countries (1950 to 2009)

*Source:* UN Population Division (2011)

The relatively recent emergence of HIV/AIDS is a significant barrier to low-income countries achieving the life expectancies of the middle-income and rich countries. Of particular concern is the extent of HIV/AIDS in Sub-Saharan African countries which had 68 per cent of all HIV cases in 2009. In 2007, 6 per cent of the population in Sub-Saharan Africa were estimated to have the HIV infection with a life expectancy nine years lower than a situation without HIV/AIDS. For some individual countries the situation is particularly severe (see Table 9.3). By some estimates Zimbabwe has a life expectancy that is 26 years lower than a situation without HIV/AIDS (Lomborg, 1998) which accounts for the fall in life expectancy that occurred in Zimbabwe during the 1990s (see Figure 9.1).

The reasons for the high levels of HIV/AIDS in Sub-Saharan Africa are many and country specific. In many countries, the stigma associated with HIV and AIDS complicates the spread of information, testing, political action and safer sexual practices that are required to tackle the epidemic. Nelson Mandela recognised and challenged this stigma in 2005 by publicly acknowledging AIDS as the cause of the death of his son:

> Let us give publicity to HIV/AIDS and not hide it, because [that is] the only way to make it appear like a normal illness. (BBC News, 2005)

Table 9.3   Prevalence of HIV amongst adults aged 15–49 in Sub-Saharan African
countries, 2001 and 2009

| Country | Adult (15–49) HIV prevalence. [Higher and lower estimates] | |
| | 2001 | 2009 |
| --- | --- | --- |
| Botswana | 26.3 [25.5 – 27.4] | 24.8 [23.8 –25.8] |
| South Africa | 17.1 [16.7 – 17.5] | 17.8 [17.2 – 18.3] |
| United Republic of Tanzania | 7.1 [6.7 – 7.7] | 5.6 [5.3 – 6.1] |
| Zambia | 14.3 [13.7 – 15.0] | 13.5 [12.8 – 14.1] |
| Zimbabwe | 23.7 [22.8 – 24.9] | 14.3 [13.4 – 15.4] |

*Source:* UNAIDS (2010)

The position of Mandela is in sharp contrast to that of Thabo Mbeki (president of South Africa 1999–2008) who at the International AIDS conference in Durban, rejected the accepted scientific wisdom, claiming AIDS was brought about by the collapse of the immune system – not because of a virus (Boseley, 2000).

In Sub-Saharan Africa the number of women with the HIV infection is higher than men. The UNAIDS report on the global AIDS epidemic estimated that there were around 12 million women living with AIDS or HIV compared to 8.2 million males. This is partly for biological reasons and the greater permeability of the vaginal mucous in comparison to the penis. However, social factors are also involved; in many developing and Sub-Saharan countries women have a relatively low status compared to men with less access to education, income and healthcare. These factors make them more susceptible to HIV infection as they lack the resources to obtain contraception or to insist on its use. The lack of economic independence and stability makes women more prone to sexual exploitation and violence as they may have to endure abusive relationships or turn to prostitution for economic survival. A final factor is the internal conflicts in countries such as the Democratic Republic of Congo which, like many wars throughout history, involved widespread sexual violence against women increasing their risk of HIV infection.

A study of mine workers in South Africa (Williams et al., 2009) examines the circumstances in which men pay for unprotected sex with local sex workers in spite of both parties having knowledge of the risks of HIV infection. This research found that the demanding and lonely nature of the mining occupation and culturally developed sexual norms of demanding skin to skin sexual contact were responsible for the miners' behaviour. For the women the risk-taking was a result of their poverty and gender discrimination.

There are grounds for optimism in relation to HIV/AIDS in Sub-Saharan Africa. The 2010 UN report on the AIDs epidemic reports that in 22 countries in Sub-Saharan

Africa, HIV incidence has fallen by more than 25 per cent between 2001 and 2009 (UN, 2010). The biggest epidemics in Sub-Saharan Africa – Ethiopia, Nigeria, South Africa, Zambia, and Zimbabwe – have either stabilized or are showing signs of decline. Furthermore, the report shows that five countries, Botswana, South Africa, United Republic of Tanzania, Zambia and Zimbabwe, showed a significant decline in HIV prevalence among young women or men in national surveys.

## Inequalities in health in the UK

The previous section examined inequalities in mortality between countries. However, it is important to recognise that inequalities also exist within countries. In the UK, health inequalities have been shown to exist between socio-economic groups and places for a number of health measures including mortality (and many specific causes of mortality), self-reported morbidity and disability, clinical measures (such as blood pressure and body mass index) as well as administrative measures (such as sickness absences and doctor consultations) (Marmot, et al., 1995; Graham, 2000; Shaw et al., 2002; Bajekal and Prescott, 2003; Bellaby, 2006). Health inequalities are an international issue and are found in all modern societies (Shaw et al., 2002; Kunst, 2005). Research suggests that health inequalities are greater for men than women (Matthews et al., 2006) and at older ages (Berney et al., 2000). We now examine the extent of social and spatial inequalities in the UK considering the theories that have been proposed to explain these health inequalities.

## Inequalities between population groups

Table 9.4 and Figures 9.2, 9.3 and 9.4 demonstrate the extent of health inequalities between different social groups using various health measures. Table 9.4,

Table 9.4   Life expectancy at birth: by social class and sex, 1997–9

| Social class | Males | Females |
| --- | --- | --- |
| Unskilled manual | 71.1 | 77.1 |
| Semi-skilled manual | 72.7 | 78.5 |
| Skilled manual | 74.7 | 79.2 |
| Skilled non-manual | 76.2 | 81.2 |
| Managerial and technical | 77.5 | 81.5 |
| Professional | 78.5 | 82.8 |

Source: Babb et al. (2004)

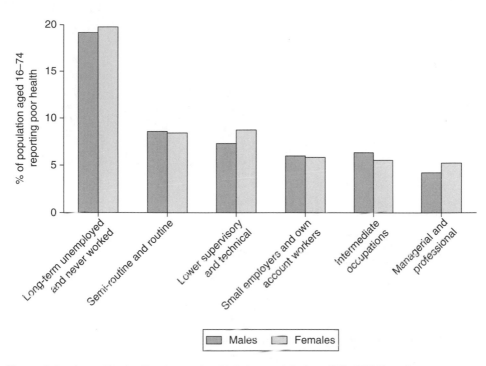

Figure 9.2    Age-standardised poor health1: by social class (NS-SEC2) and sex,
2001, UK

*Source:* Office for National Statistics – Focus on Social Inequalities Report

which is based on data produced by the Office for National Statistics for the years 1997 and 1999, reveals that men working in professional occupations could expect to live 7 years longer than those who worked in unskilled manual occupations. Figure 9.2 shows that rates of self-assessed poor health are almost twice as high in semi-routine and routine occupations compared to professional and managerial occupations as recorded in the 2001 Census. Additionally, one in five of those who are long-term unemployed or who have never worked are in poor health – four times the rate for managers and professionals. We also observe important differentials by ethnicity (Figure 9.3, on page 186); age-standardised census rates (2001) of limiting long-term illness are highest amongst Pakistani men (14 per cent) and women (17 per cent) and lowest for Chinese men (5.6 per cent) and women (6.2 per cent). Finally, these differentials are observable at the very earliest stages of life; the rate of infant mortality for sole-registered babies (where only the mother is named on the birth certificate) and for children whose fathers are in semi-routine or routine occupations was equal to 7 per 1,000 live births in

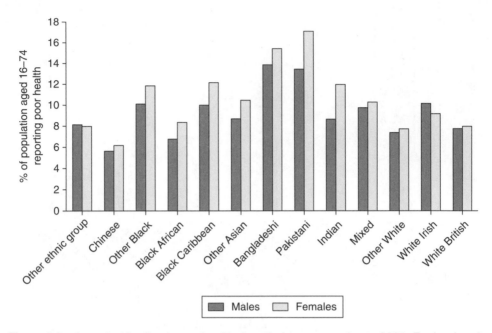

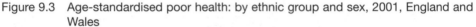

Figure 9.3   Age-standardised poor health: by ethnic group and sex, 2001, England and
            Wales

*Source:* Office for National Statistics – Focus on Social Inequalities Report
*Note:* Poor health defined as 'not good' health in the last 12 months for persons aged 16–74

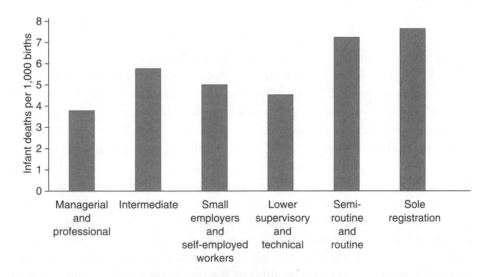

Figure 9.4   Infant mortality rate by registration type and father's social class (NS-SEC):
            2000–02, England and Wales

*Source:* Office for National Statistics – Focus on Social Inequalities Report

2000–2002, twice that of children born to fathers in professional or managerial occupations (Figure 9.4, overleaf ).

Mel Bartley (2004) provides an excellent and detailed account of the main theories that have been developed to explain the health inequalities between socio-economic groups. Here we summarise the six main models that she identifies:

## Materialist model

In the UK a key document in the explanation of health inequalities is the Black report (published in 1980 by the Department of Health and Social Security) which formally identified a link between social and economic deprivation and health inequalities attributing the widening health gap to a materialist model of health inequalities. This theory proposes that the low income received by those in the lower social classes leads to them experiencing less favourable conditions in a number of areas such as working conditions, quality of local environment and housing, diet, access to health services/information and education which results in inequalities in health between social classes (Bartley, 2004) The findings of this report initially met resistance – *The Health of the Nation* white paper, published by the UK Government in 1992, refused to accept that social and economic factors could influence the health of individuals (Guy, 1996). However, increasing evidence has emerged so that now there is little doubt that deprivation has an impact on health, although the reasons for this link are disputed (Staines, 1999).

## Selection model

The selection theory has parallels with Darwin's theory of natural selection proposing that the characteristics of each individual contain different potential for good health. The theory proposes that those with the traits associated with good health (e.g. intelligence, energy) gravitate towards the higher social classes and so those in higher classes tend to have the better health than those in lower classes. Longitudinal data (where information on individuals is available at several time points) has proved very important in assessing the merits of this particular model. Research using the Longitudinal Study for England and Wales (a sub-sample of the census that is linked over time) reveals two weaknesses of this model. First, researchers demonstrated that health inequalities were almost unaffected after they removed the social mobility from the longitudinal data (anyone who had moved from one social class to another was removed from their analysis). Second, most movement between social classes occurs at the youngest working ages and as mortality and morbidity rates are lowest at these ages it is unlikely that this social mobility could be very important in accounting for the health inequalities we observe (Guy, 1996).

## Cultural/behavioural model

The cultural/behavioural theory attributes differences in the health of people across social classes to the prevalence of unhealthy lifestyle choices within these groups (Black et al., 1982). Smoking, one of the greatest single causes of preventable illness and premature death in the UK, is often suggested as a key pathway in which the actions of the less well-off lead to health inequalities. In 2009, smoking prevalence was around three times higher among those from the lowest income households (40 per cent for men and 34 per cent for women) than among those living in the highest income households (14 per cent for men and 11 per cent for women) (Wardle, 2009).

It is important to recognise that not all unhealthy lifestyle choices are most prevalent in the lower social classes. For example, heavy alcohol consumption has been linked to a number of medical conditions such as mouth, throat, stomach, liver and breast cancer, high blood pressure, cirrhosis and depression and is most prevalent in the higher social classes. Analysis of the 2009 Health Survey for England reveals that those with higher incomes were more likely to have drunk more than twice the recommended limit in the last week whilst the maximum daily alcohol consumption was similar across income groups (Fuller, 2010). Similarly, whilst there is evidence that levels of obesity are higher in lower income groups compared to higher income groups for women, no such pattern is apparent for men (Tabassum, 2010).

The cultural/behavioural theory has some links with the material explanation in the Black Report because unhealthy lifestyle choices, such as smoking, are likely to be influenced by the limited opportunities and financial restrictions experienced in low income groups (Staines, 1999). However, research suggests the cultural/behavioural model cannot explain all of the health inequalities we observe between social groups. For example, the Whitehall studies of health inequalities amongst British civil servants found that differences in health related behaviour explained about a quarter of the differences in mortality risks between the grades of civil servants (Bartley, 2004).

The ideas of personal responsibility contained within this model certainly accorded with those held by the UK Conservative government of the 1980s who preferred this explanation to the materialist model adopted by the Black Report (Guy, 1996). However, as Guy (1996) notes, Labour politicians in the 1970s also argued for greater individual responsibility in terms of health. The idea of the 'nudge' (discussed in the Epidemiological Transition Model section of Chapter 2) whereby people are encouraged to adopt more healthy lifestyle choices is now a central part of health policy proposals of all of the major political parties:

> Today we can't escape the fact that today many of our most severe health problems are caused, in part, by the wrong personal choices. Obesity, binge-drinking, smoking and drug addiction are putting millions of lives at risk and costing our health services billions a year. So getting to grips with them requires an altogether different approach to the one we've seen before. We need to promote more responsible behaviour and encourage people to make the right choices about what they eat, drink and do in their leisure time. (David Cameron, foreword, A Healthier Nation, Policy Green Paper No. 12, Conservative Party)

## Psycho-social model

The psycho-social theory of health inequalities emerged from suspicions that the health differences between advantaged and disadvantaged groups were too large to be explained purely by material disadvantages stemming from differences in income (Guy, 1996). This theory relates the psychological effects of stressful conditions at work, home, or that result from low social status, to physiological effects on health (Bartley, 2004). The Whitehall studies are particularly important pieces of evidence for this theory of health inequality. The Whitehall I study was set up in 1967 and followed the health of male civil servants over time focusing on heart disease and other chronic conditions. It was thought that the stress linked to managerial and executive occupations would lead to higher mortality rates in these occupations. The research actually revealed the opposite; men in the lowest grades had a death rate three times higher than in the highest grades. The White-hall II study revealed that low job status was also not only responsible for higher levels of heart disease but also similar patterns were evident for some cancers, chronic lung disease, gastrointestinal disease, depression, suicide, sickness absence from work, back pain and self-reported health (Wilkinson and Pickett, 2009). Research by Wilkinson (1997) suggests that the psychological effects of social position account for the larger part of health inequalities compared to the effects of material circumstances.

## Life course model

The life course model of health inequality has emerged in the 1990s with increasing availability of longitudinal data on health status. The model proposes that chances of good or bad health are influenced by life events including in very early childhood and even before birth (Berney et al., 2000). A key strand of research within life course studies is the examination of how key events in the life course affect health and health inequalities at older ages. For example, Westerlund et al. (2009) investigate the impact of retirement on self-assessed health using longitudinal data (throughout retirement) on employees of the French national gas and electricity company. The research finds that the burden of ill-health, in terms of perceived health problems, is substantially relieved by retirement for workers in poor working conditions (often found in low-grade occupations) with no such effect observed for those in ideal working conditions (most common in high-grade occupations).

## Neo-materialist model

The neo-materialist theory examines the way in which societies as a whole differ according to social, political and cultural factors and the impact that these differences have on health between nations (Whitehead et al., 2000; Bartley, 2004). A neo-materialist approach to health inequalities taken by Wilkinson and Pickett (2009) is described later in this chapter.

The theories that are discussed here are not mutually exclusive and it is likely that several theories combine to explain the processes leading to health inequalities in society (Bartley, 2004). For example, material, behavioural and psycho-social factors cluster together; those in the lower social classes are likely to be exposed to the health risks associated with all three theories across the course of their life-time (Graham, 2000).

## Inequalities between areas

Research demonstrates that health inequalities exist between areas within the UK. Figure 9.5 shows standardised illness ratios across all UK districts using the 2001 Census measure of limiting long term illness (LLTI). An SIR above 1 indicates that an area has a higher than average level of LLTI after its age structure is taken into account. Districts are grouped into five equal groups on the basis of their SIR. The lowest SIRs are predominantly found in Southern England and around London whilst the highest SIRs correspond to former manufacturing and industrial and urban areas in the North of England, South Wales and Scotland.

The extent of deprivation within an area also turns out to be a good indicator of population health. Figure 9.6 shows the inequalities in life expectancy and healthy

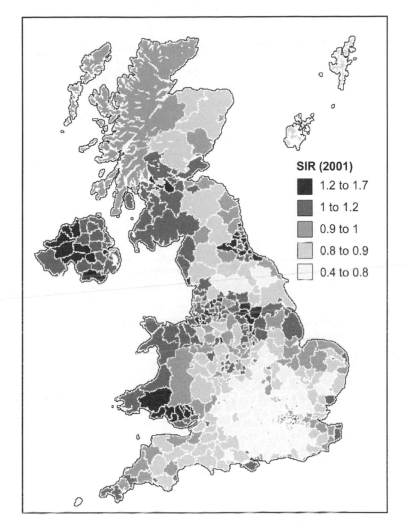

Figure 9.5   Map of standardised illness ratios for UK Districts, 2001

*Source:* ONS area classification and authors' own calculations using data from the 2001 Census (Tables ST16 and ST65)

life expectancy (the number of years a person can expect to live in a healthy state) between the most and least deprived wards (Bajekal, 2005). In this research, each ward in the UK was given a deprivation score using the Carstairs deprivation meas ure which is an unweighted combination of four indicators of material deprivation – namely, the proportions of people in households headed by a person in a semi-skilled or unskilled manual occupation (Social Class IV or V); economically active men

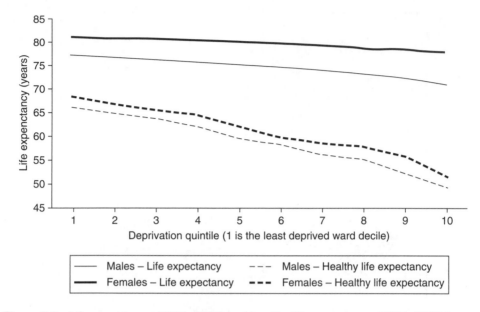

Figure 9.6   Life expectancy (1994–1999) and healthy life expectancy (1994–1999) by
ward deprivation decile and sex

*Source:* Bajekal (2005)

seeking work; persons with no car; and persons living in overcrowded accommodation. Wards are ordered according to their deprivation score (least to most deprived)
and then divided into 10 equal groups (deciles). Healthy life expectancy is calculated
using Sullivan's method (see Box 9.2) applied to data from the 'general health' question (see Box 9.1) in the Health Survey for England. As we move from the least
deprived to the most deprived wards, life expectancies fall by around six years for
men and three years for women. A man living in the most deprived areas of England
has a life expectancy of 71 years which is similar to that experienced by the average
person in countries such as Brazil, China and Libya in 2001. An even stronger social
gradient is found for healthy life expectancy with both men and women in the least
deprived wards enjoying almost 17 additional years of good health compared to
those in the most deprived wards.

The spatial patterns of mortality and ill health are remarkably stable over time
(Graham, 2000). Not only have these spatial inequalities persisted over time but
research suggests that they may have expanded over the last 30 years for a range of
mortality causes and health indicators (Marshall, 2009; Shaw et al., 1999; Davey
Smith et al., 2002).

One potential explanation for the polarising patterns of health links such changes to
the parallel expansion of social and spatial inequalities in income, wealth and poverty

that leads to increasing health inequalities through the material/structural and psycho-social theories of inequality (Shaw, 1999).

A second explanation is that patterns of migration serve to exacerbate the existing spatial patterns of inequalities in mortality (Brimblecombe et al., 1999) and morbidity (Norman et al., 2005). As discussed in Chapter 6, migrants are more likely to be in the higher social classes, to have jobs and be healthy. Research demonstrates a negative relationship between area population change and the level of mortality, indicating that those who are financially and physically-able leave areas of deprivation for more healthy areas (Davey Smith et al., 1998; Dorling et al., 2000). Conversely for those in affluent healthy areas who become ill, a move to a more deprived area may be forced by a job loss or change of occupation and inability to afford rent or mortgage repayments (Dorling and Thomas, 2004).

A third explanation for spatial polarisation of health concerns the changes in the socio-economic composition of housing tenure and the spatial polarisation of property values. The rapid rise in home ownership and decline in the local authority housing stock has led to housing tenure (and neighbourhoods) being increasingly divided according to socio-economic status. In addition, the spatial polarisation in wealth in the UK that occurred since the 1970s is also visible in terms of property wealth, with the gap between property prices in the most and least desirable places growing (Dorling et al , 2000). Those in rented accommodation and who live in the cheapest accommodation have lower life expectancies and are more likely to be in poor health than owner occupiers and those owning expensive properties (Dorling et al., 2000; Macintyre et al., 2000). The association between tenure/property value and health is partly because poorer people who are more likely to have health problems are also more likely to rent or live in cheap property. However, living in rented/cheap accommodation can also cause poor health through physical/social features of the home and area such as damp accommodation and/or area deprivation (Best, 1999).

## Relationship between health and locality

Research suggests that health inequalities between areas are caused by both compositional (characteristics of the individuals in an area) and contextual (characteristics of the area itself) factors (Gould and Jones, 1996) with the relative contribution of each being subject to debate.

Contextual (area) factors that might influence health after accounting for the composition of an area include the quality of local environment, deindustrialisation and associated economic decline, housing quality and tenure, the presence of polluting industry, the food on offer in local shops, quality of local services, levels of social capital, sense of community, crime (and the fear of crime) and the provision of sporting and leisure facilities (Dorling et al., 2000; Graham et al., 2000; Joshi et al., 2000). From a policy perspective, the relative importance of contextual and compositional explanations of health inequalities between places can lead to a tension between the pursuit of policies that are aimed at areas and those that are aimed at individuals.

Shaw et al. (2002) consider several pieces of research that examine the roles of contextual and compositional factors in determining the health of the population in particular areas. An interesting piece of work in their review that illustrates the importance of area specific or contextual factors in determining the levels of health in an area is the comparison of premature mortality in Middlesbrough and Sunderland (Phillimore and Morris, 1991). This study shows that despite having very similar socio-economic characteristics, levels of premature mortality were consistently and markedly higher in Middlesbrough in the early 1980s. One difference between the two areas is that whilst both have consistent high levels of unemployment, Middlesbrough has experienced fluctuations between periods of 'boom and bust' in terms of levels of employment. It is suggested that the decline into deprivation during bust periods is particularly harmful to health.

## Wealth inequality and health inequality

An interesting observation is that as countries reach a certain level of wealth (GDP per capita above US$5,000) further increases in wealth have less effect on life expectancy and for the richest countries the relationship between increasing wealth and life expectancy breaks down completely (Wilkinson and Pickett, 2009).

In *The Spirit Level: Why Equality is Better for Everyone*, Richard Wilkinson and Kate Pickett take a neo-materialist approach arguing that for the richest countries the level of income inequality within the country predicts the level of population health according to a number of different health indicators (life expectancy, infant mortality, birth weight, AIDS and depression). Their hypothesis is that the more evenly a country's wealth is spread, the better the health of that country is. The book displays a number of graphs that provide evidence to support this hypothesis through the association between these health measures and the level of income inequality (as measured by the ratio of the income received by the top 20 per cent to the bottom 20 per cent). Wilkinson and Pickett explain their finding using the psychosocial theory for health inequalities (discussed earlier in this chapter). They claim that people's position in society causes stress which influences health beyond explanations provided by alternative theories such as smoking or other unhealthy lifestyle choices. They attribute the declines in life expectancy in Russia to the move from a centrally planned to market economy which resulted in an explosion in extent of income inequality.

The theories put forward in *The Spirit Level* have stimulated a great deal of interest. For example, both David Cameron and Ed Miliband, the leaders of the two main political parties in the UK at the time of writing, have endorsed its messages. At the same time, a number of researchers have challenged the Spirit Level hypothesis, for example, Peter Snowdon, the author of *The Spirit Level Delusion*, argues that *The Spirit Level* does not sufficiently account for other factors that differ between countries which might undermine the hypothesis they present. Additionally, Snowdon claims that the exclusion of certain countries undermines the associations between income inequality and the extent of societal problems. Wilkinson and Pickett respond to these criticisms by pointing out that countries were selected according to a strict set of rules with no departures or exceptions. They note that their work presents a theory of problems which display a social gradient rather than a theory of everything and that they do not claim that income inequality is the *only* cause of worse social or health problems in a society.

## Summary

- The study of population health is important to quantify the extent and development of health inequalities between and within countries and to inform strategies to tackle these inequalities and to improve population health more generally.

- Health is a contested concept and a number of definitions can be used. For example, we could define health through the absence of a medically identifiable illness or condition. Alternatively, our definition might focus on the ability of individuals to perform everyday activities or an individual's own self-assessment of their health. Each definition gives different levels of poor health in a population. Estimates of ill-health are also affected by societal understandings of what health is which change over time and between places.

- A number of data sources are used for health research. Most countries record mortality statistics and many undertake censuses or surveys that include self-assessed health measures and in some cases clinical measures of health. Administrative statistics on health include GP consultations, hospital visits and benefits data but these only count those who have health problems and access services.

- There have been dramatic improvements in life expectancy over the last 100 years. In 1900, world life expectancy was around 30 years. By 2008, more than 85 per cent of the world's population could expect to live to at least the age of 60. Sub-Saharan African countries lag behind the rest of the world in terms of experiencing a very modest increase in life expectancy over the last 50 years. In some countries, such as Zimbabwe, life expectancy fell between 1950 and 2008. AIDS is an important factor in explaining the lack of improvement in levels of mortality in Sub-Saharan Africa.

- In the UK there are inequalities according to a number of health indicators between population groups based on social class, income and ethnicity. A number of theories have been developed to explain these inequalities. The theories are not mutually exclusive; material, behavioural and psychosocial factors act together and are experienced disproportionately in the lower social classes throughout their lifetime.

- Strong spatial health inequalities exist in the UK and it is thought that these are caused by compositional (individual) and contextual (area) factors. The poorest health is found in the urban and former industrial and manufacturing areas of the North of England, South Wales and Scotland. The best health is found in the affluent districts that surround London.

- Often debates around health inequalities have political dimensions. For example, the arguments put forward by Richard Wilkinson and Kate Pickett, that more equal countries tend to have better health outcomes, have attracted considerable interest as well as scrutiny and challenge.

## Recommended reading

For a detailed discussion on concepts of health see Michael Senior and Bruce Viveash (1998) *Health and Illness*. Basingstoke: Macmillan Press. Those interested in an evaluation of various sources of data on health should refer to John Charlton's chapter, 'ONS data: other health sources', in Diana Leadbetter (ed.) (2000) *Harnessing Official Statistics*. Abingdon: Radcliffe Medical Press.

Gattrell and Elliott (2009) *Geographies of Health: An Introduction*. Oxford: Blackwell, provide more detailed information on health inequalities on a global scale. More details on this and the issues of AIDS and HIV with a focus on developing countries can be found in Glyn Williams, Paula Meth and Katie Willis (2009) *Geographies of Developing Areas: The Global South in a Changing World*. Abingdon: Routledge.

Mel Bartley (2004) *Health Inequality: An Introduction to Theories, Concepts and Methods*. Cambridge: Polity Press, and Hillary Graham (2001) *Understanding Health Inequalities*. Buckingham: Open University Press, provide a detailed discussion of theory on and extent of the health inequalities in the UK. For a focus on health inequalities between geographical areas, see Mary Shaw, Daniel Dorling and Richard Mitchell (2002) *Health Space and Place*. Harlow: Pearson.

For a flavour of the debate around research on health inequalities see Kate Pickett and Richard Wilkinson (2009) *The Spirit Level:Why More Equal Societies Almost Always do Better*. Cambridge: Polity Press, and Peter Snowdon (2010) *The Spirit Level Delusion: Fact Checking the Left's New Theory for Everything*. London: Little Dice.

# 10

# POPULATION FUTURES

In this final chapter we consider some of the key issues that are likely to shape population and population research in the future. Our intention is not to provide a forecast of population dynamics or of policy, or to provide an exhaustive list of current population research, but rather to consider how current issues have evolved out of past debates as well as population processes. This chapter therefore focuses on two main issues – the possible threat of overpopulation and the realities of an ageing population. In considering these two issues, we attend to our three themes: the significance of scale; the magnitude of inequality; and the importance of data and their interpretation. Finally, we consider how population research itself might be shaped in the future.

## Overpopulation?

A cursory glance at media coverage of UNFPA's *State of the World Population 2011* published in October of that year with the special theme of 7 billion people, would conclude that future global population growth is unsustainable and will be damaging for the global environment. Yet while journalists and commentators might be in broad agreement about the need to control population for 'the sake of the planet', among population scientists the impact of future population growth is contested. We have already considered some of these debates, for example the infamous Ehrlich and Simon bet of 1980, (see Chapter 1); in this section we review the broad arguments on either side of the debate.

Increasing global population is often perceived as problematic for a number of reasons. Population growth has been blamed for the lack of economic development in Third World countries, inter and intra-national conflict, out-migration, depletion of the Earth's resources, environmental degradation and more recently global

warming (Connolly, 2008). Theoretically academic interest in the perils of population growth has sought to re-engage with Malthusian interpretations of population
growth (see Chapter 2). A bibliographic search on Malthus will reveal various publications that deal with 'his return' or a 'reassessment' of his work. Why then is a reassessment of Malthus appropriate and why are his ideas returning to influence debates
on population?

One response is quite simple; Malthus was broadly accurate for his time but wrong
about future population growth as he did not predict the technological, social and
economic advances of the nineteenth century. Yet his writings about population rather
than rejected are reinterpreted for different economic, social, cultural and environmental contexts. Malthus' assertion that the relationship between population and
resources is fixed is the main flaw in his theory, and one that quickly unravelled in the
nineteenth century. Karl Marx dismissed Malthus as an advocate of bourgeois society
who failed to recognise the true plight of the working classes under capitalism. Though
in other areas his ideas were broadly accepted, most famously both Charles Darwin
and Alfred Russel Wallace were influenced by Malthus's writings in formulating the
theory of natural selection. Yet despite the fact that history has not been favourable to
Malthus, his theories have survived as an influential contribution to population
debates. Neo-Malthusian interpretations emerged in the 1960s and 1970s, associated
with a number of key American contributors including Paul Ehrlich (1968) and Lester
Brown (1988). Ehrlich in particular is associated with outlining the limits to population growth and in making the case for preventative checks on population growth. In
1968 the global think-tank the Club of Rome (COR) was formed to, amongst other
aims, 'stimulate general debates on the major issues that have global implications for
all aspects of the human condition, taking a holistic approach that covers their moral,
material, cultural, social and scientific aspects' (Club of Rome, no date). In 1974, the
COR published *The Limits to Growth* (Meadows, 1974) that considered the implications
of rapidly growing world population and finite resources. The apparent need for constraint in population growth has not though been universally accepted; especially in the
developing world where commentators and activists argue that population growth is
the result of poverty, and not its cause as Malthusian theory predicts (Gould, 2009:
54–6). Thus some population scientists and economists have argued that population
growth is a necessary factor for economic prosperity and technological advancement.
Furedi (1997) highlights the paucity of evidence to support the theory that population
growth inhibits development in poor countries. Simon (1982) challenges the idea of
resources as 'fixed' and 'knowable' suggesting that new resources are constantly being
discovered and utilised. In his book *The Ultimate Resource*, while sharing Malthus' view
of the connection between population and economic growth, Simon turns the relationship round so that it is a positive one; individuals themselves are the resource that
drives wealth creation. Lomborg (2001) points out that over the previous century our
utilisation of resources has surpassed population growth and supplies do not appear to

be in danger of running out in the immediate future. The Danish economist Ester Boserup argued in the 1960s that population growth stimulates the need to enhance agricultural yields and find technological solutions (see Gould, 2009: 63–70 for a review of Boserup and Malthus).

Yet Malthus' revival in the twentieth century has a twist in the tale, in that his preoccupation with agricultural productivity has been replaced by a concern about 'the environment'. Population growth in the twenty-first century is popularly believed to be outstripping environmental resources, and moreover that it will bring about further environmental damage. In the UK, both the Royal Commission on Environmental Pollution and the Royal Society have investigated the relationship between population and the environment. Overpopulation is a key current concern as illustrated by Jeffrey Sachs in his 2007 Reith Lecture:

> Our planet is crowded to an unprecedented degree. It is bursting at the seams. It's bursting at the seams in human terms, in economic terms, and in ecological terms. This is our greatest challenge: learning to live in a crowded and interconnected world that is creating unprecedented pressures on human society and on the physical environment. (Sachs, 2007)

Furedi's (1997) review of late-twentieth-century neo-Malthusian thought outlines how focus has transferred from concerns about the ability of populations to feed themselves; to a concern that overpopulation leads to environmental degradation. Furedi concludes that the link between population growth and environmental damage is contested because it does not adequately account for factors such as technology that reduce environmental impact or for the capacity of people to improve their environments. Though environmental neo-Malthusianism might be appealing and 'common sense', empirically it is less well grounded. Moreover, while the relationship between population growth and environmental degradation is popularly conceptualised at a global level, it is not empirically possible to demonstrate the causality of the relationship at this scale. At smaller scales – regional, national or subnational – the evidence for the relationship between population growth and either poverty and/or environmental degradation, is less conclusive. There are parts of the world where the Malthusian trap seems inevitable if we apply the (unrealistic) assumption that a country must be able to feed itself; in parts of West Africa there is limited potential for increased agricultural productivity to feed a growing population. Thus Niger, which recorded the highest global fertility rate of 7.1 in 2009, is reliant on food aid to mitigate the risk of famine. Agricultural yields in Niger, as in other parts of the Sahel, have been badly affected by excessive heat and drought. In these fragile environments, the logic of Malthusianism may be more certain (Gould, 2009: 58). In contrast, there are case studies of the how abandonment of farm land and associated population loss can cause environmental degradation. For example, Tiffen et al. (1994) in a very influential, yet also controversial, study of the Machakos region in Kenya, called *More People, Less Erosion* documents how population growth has been sustained and moreover has brought about social and economic conditions that are triggers for environmental improvements, such as soil improvements and terracing of slopes.

The debates on population, development and the environment are important because they have the potential to influence policies that manage human impact on the environment and address economic inequality. If population growth is viewed as an environmental problem, this might imply the need for policies that attempt to reduce fertility. On the other hand, to the extent that population growth is not the cause of environmental or social problems, then there is a need to conduct research to inform more appropriate policies. There is considerable popular support for a Malthusian interpretation of the relationship between environmental resources and population growth. In particular, the idea that environmental resources are fixed and can only be degraded further by population growth has become a truism in the twenty-first century and one for which there appear to be few dissenting voices, or these voices are not easily heard. Yet we could also conclude that since writing his essays, the relationship between population and resources has become far more complex than that originally outlined by Malthus. His ideas are appealing because of their simplicity. However neo-Malthusianism does not necessarily consider the dynamics of the relationship between population and resources, namely population growth can stimulate technological innovation for more efficient use of resources and discovery of new resources and does not necessarily have to bring about resource degradation. A more cautious and balanced evaluation is that geography matters. The relationship between population and environment has to be put in context and the impact of locality should not be ignored (Malakoff, 2011). Locality matters because of both physical and social differences, as the kinds of resources available are contingent on locality, but so is access to urban markets and technological progress. The challenge is to highlight localities, such as the Machakos, that illustrate how Boserup's view on the relationship between population and resources can be brought about. We should not assume an automatic and given relationship between population and resources.

# Ageing or inter-generational relations?

The idea of population ageing as inevitable and with ominous implications is now widely felt and is a major anxiety of our current time (Mullan, 2000). The process of population ageing involves an increasing proportion of the population at the oldest ages as a result of falling fertility rates and increasing life expectancies. The process of population ageing is set to continue throughout the twenty-first century but at different rates in the various regions of the world. In developed regions, the proportion aged over 65 is projected to increase from 14.3 per cent to 26 per cent between 2000 and 2050. In developing regions, the projected increase is from 5 per cent to 14 per cent (McCracken and Phillips, 2005).

Population ageing can be regarded as a success story. Improvements in health, nutrition and wealth mean that more people are surviving to the older ages. However, for many, population ageing represents a cause for concern. It is claimed that the growth of the elderly population will lead to increases in the population with disabilities or poor health who will require some form of support (either financial, medical or care) that a diminished working-age population will be less able to provide (Thane, 1989). For some analysts, the rising costs associated with population ageing, most notably stemming from healthcare provision and pensions, can only be met through increasing the tax burden on the working age population (Bos and Weizsacker, 1989). Attention focuses on the growth of the elderly population because they have greater need for medical and care services compared with other ages (Jarvis, 2000) and healthcare costs are greatest at the older ages (Jouvenal, 1989; McCracken and Phillips, 2005). In the UK, population ageing has become an increasingly high profile subject in the news and print media. The BBC chose population ageing as an issue in the 'If' series that explored and analysed the big issues facing the UK in the future (BBC, 2004).

However, a number of arguments have emerged that are more optimistic about both the scale of the challenge and the implications of population ageing. It is important, first of all, to note that population ageing is not a new phenomenon nor a permanent one. In the UK in 1891 there were 12 working age people for every person aged over 64 compared to four working age people for each person aged over 65 in 2001. Society coped adequately with the population ageing over this period, with increased support to the elderly through state-funded pensions, health provision and many other services, partly because increases in productivity outstripped the growth of retirees. Furthermore, population ageing is not expected to continue indefinitely; population projections suggest that that the old-age dependency ratio will level out by the middle of this century.

Focusing on the older population ignores changes expected at other ages. For example, growing government expenditure to support retired people can be balanced against support for reduced numbers of children. The working age population is often considered a homogenous group of 'contributors' in debates on population ageing without consideration of factors such as unemployment which can have a significant impact on the welfare budget of a government.

Most elderly people live independently and contribute to society, often in unheralded ways, through voluntary work, helping out with family childcare and personal consumption (Freer, 1988; Bond and Rodriguez Cabrero, 2007). Although rising care costs over time are regularly attributed to population ageing there are other factors involved such as new medical technologies and rising consumer demand that are major expenditure drivers (Anderson and Hussey, 2000; Evans et al., 2001; McCracken and Phillips, 2005). Scherbov and Sanderson (2011) develop new measures of population ageing that take improvements in longevity and health into account. They argue that traditional measures (such as dependency ratios) exaggerate the extent and challenges of population ageing projected for the future. Research by Zweifel et al. (1999) shows that age is actually a weak predictor of healthcare expenditure, once time to death is controlled for suggesting that increased life expectancy does not increase the overall health support needed by each person. Mullan (2000) claims that the pace of economic expansion projected into the next half century, although not guaranteed, is sufficient to cover the extra societal costs associated with population ageing.

Thus in the developed world, policy makers are primarily concerned with the problems of dealing with an ageing population. In many developing countries, the main population issues are more closely related to growth and the challenges of feeding, housing and schooling an increasing population, although these populations are also ageing as countries complete the demographic transition. Therefore, it is too simplistic to reduce future population challenges to a concern about growth or ageing, rather the challenges of inequality dominate. Inequality relates to differentials in growth, but also to the balance between population and resources and the challenges of changing population structures. The UNFPA *State of the World's Population 2011* which took as its special theme a global population of 7 billion, focused not just on growth but on the challenges of finding opportunities for global youth in developing countries, where young people are faced with the very real problem of being crowded out of the labour market. Young people and adolescents made up nearly 2 billion of the global 7 billion people in 2011, and it is this age group that is integral to future population growth. The UNFPA 2011 Report states that providing education opportunities, promoting gender equality and access to family planning services are essential in order for the current global youth to complete the global fertility transition.

# Demographic solutions?

It is interesting to consider if there are demographic solutions to the potential problem of overpopulation and the reality of population ageing. Both issues are concerned with unevenness between population and resources, though in different ways. Might population growth in some parts of the world provide the solution to ageing populations elsewhere through migration? This is clearly a contentious proposal. In many developed economies, governments are attempting to reduce migration as it is seen as a threat to national prosperity and the work and housing needs of the local-born population. The

Table 10.1  Projections for replacement migration in Japan

| Year | Projected population: no migration (thousands) | 1995–2050 | | |
|------|------|------|------|------|
| | | Number of migrants to maintain constant population (thousands) | Number of migrants to maintain constant population aged 15–64 (thousands) | Number of migrants to maintain constant ratio of 15–64/65 years or older (thousands) |
| 1995 | 125472 | | | |
| 2000 | 126714 | | | |
| 2025 | 121150 | | | |
| 2050 | 104921 | 17141 | 33487 | 553495 |

*Source:* Table 21 Population indicators for Japan by period for each scenario, http://www.un.org/esa/population/publications/ReplMigED/Japan.pdf

limitations of migration to solving demographic challenges are illustrated in a UN report published in 2000 on replacement migration. This report considered how migration could alleviate the problems of an ageing population in selected developed countries. Yet in Japan, as Table 10.1 illustrates, the number of migrants needed to alleviate the demographic challenge of ageing is considerable. For example, 500 million migrants would be required to sustain the ratio of 15–64 year olds to over 65 year olds at the same level in 2050 as it was in 1995. This figure is more than four times as large as the population in Japan in 2000. It would seem reasonable to conclude that replacement migration cannot maintain current population age structures on its own as the numbers involved are politically and practically just too big. Though some countries have seen migration on this scale, for example Singapore has actively sought to increase its population through immigration (Yeoh, 2007) and if it repeats the level of net migration in 2010 Singapore would require 28 years to quadruple its population.

Yet while international migration is a significant challenge for the developed world, and does not necessarily offer a solution to discrepancies in population structures, at different scales migration can offer a solution to the problems of balancing population and resources. In particular, urbanisation is increasingly seen as a key process through which the demographic transition may be achieved in Sub-Saharan Africa (Dyson, 2011). In one sense, this turns conventional views about the dangers of population growth on its head, as from a privileged Western view increasing population density is often viewed as one of the main problems of population growth, but it can also be a solution. In the developing world, urbanisation is associated with achieving fertility transitions and tackling childhood mortality, moreover urban living can be 'greener' living if it allows residents to develop more sustainable lifestyles.

We can conclude that population issues cannot be resolved by demographic solutions on their own. The solution to unsustainable population growth cannot just focus on reducing fertility, but also needs to address why population growth is unsustainable. It follows that this requires us to consider carefully the relationship between population and resource allocation. Likewise there is nothing that societies can do to turn back the clock of demographic ageing, as it is brought about by previous declines in fertility and mortality. Moreover population issues should not be treated as exogenous to political, economic, social and environmental challenges. The connection between population and the environment might be more transparent than the importance of other domains, but these interconnections need to be considered. How demographic processes interact with political, economic and social dimensions has implications for equality and the distribution of resources. For example, in the popular uprisings in 2011 in Tunisia, Egypt, Libya and Syria known as the Arab Spring, some commentators referred to the importance of a youthful population in these countries as a causal factor for why these uprisings occurred. These countries had a large proportion of the population in their twenties and thirties which resulted in a lot of young people with relatively few opportunities who were willing to challenge existing political regimes (Hvistendahl, 2011). Demographic structures can therefore be linked to political structures and stability, although many other factors are involved.

## Future of population studies

The advance of population research throughout the twentieth century was dependent on both the availability of data and development of techniques for the reliable and systematic analysis of demographic data. Moreover the commitment of UN agencies and other NGOs to the amelioration of demographic inequalities has stimulated demographic research but has tended to focus analysis at country or world-region level. In the twenty-first century, the problems of demographic inequality have not dissipated and population issues remain centre stage. Yet the focus of population research is also shifting.

First the formulation of demographic theory in the latter half of the twentieth century has paid less attention to differentials in scale and the significance of migration. This is exemplified in the original formulation of the DTM. Yet the development of population geography offers a different perspective, and increasingly researchers seek to reveal the dynamics of subnational population processes. The expectation of a universal theory of demographic change that is applicable across all contexts is harder to sustain. Thus researchers are more interested in understanding regional variations in population dynamics.

Research on the demographic processes of small areas relies on the availability of data, and this is one area where we can expect important changes in the first decades of the twenty-first century. In particular, the role of the census in providing key population data is being reconsidered. Some developed countries (the Netherlands, Germany, Finland) no longer carry out a decennial census and instead have introduced population registers (the Netherlands), rolling census (France) or are seeking to make linkages from administrative data. In some ways, these methods can provide more accurate and up-to-the minute population counts, yet might provide less socio-demographic detail than conventional censuses. The use of administrative data is likely to be of growing importance, particularly as methods for data linkage improve. Data is collected about ourselves in many different ways. When going shopping, filling tax returns and browsing the internet we generate data that can be captured either by private companies or the state. If this data can be harmonised and linked, then the potential scope for research is considerable. One way in which the availability of administrative data and the requirement for data at a finer geographical scale is being realised is through the increased use of geo-demographic profiles, such as those used to illustrate age structure in Chapter 4. Some commentators have argued that in the future this kind of data and spatial analysis will become more important than conventional individual-level analysis that has dominated social science research (Savage and Burrows, 2009). At the very least some of the conventions that dominated population research in the last century, particularly the reliance on nations or global regions as units of analysis, are likely to be augmented by greater attention to scale and geography.

Throughout this book we have touched upon some of the most pressing issues in contemporary societies, from population growth, ageing, migration and mobility, health inequalities, increased propensity to live alone, declines in marriage and increase in divorce. Some of these issues will remain at the core of population research well into this century, and others may be surpassed by other issues. Yet as we have explored throughout this book, our ability to make sense of population issues depends on an appreciation of the importance of data and knowledge of appropriate methodologies for analysis, awareness of the importance of scale and geography and a commitment to unmasking inequalities. Moreover, we cannot research population issues in isolation, rather population touches on all branches of social science; it is relevant for geographers, sociologists, economists, environmental scientists and political scientists. An appreciation of population issues provides a foundation for the study of individuals and society.

This book has reviewed how population researchers have understood the relationship between populations and societies. We have considered how the study of population has advanced since the late eighteenth century, with reference to both theoretical and empirical advances and we have outlined some of the possible questions that researchers might turn their attention to in the future. Throughout the book we have highlighted the importance of population dynamics and their relevance to some of the most pressing social, environmental, political and economic issues. Moreover as population geographers, we have examined the significance of scale and how geographical and temporal frameworks shape how we collect data and its interpretation.

# Recommended reading

In the year 2011 there were a number of publications to mark the birth of the seventh billion person. In particular a special edition of *Science* published on 29 July 2011 (volume 333) includes articles on population projections and the challenges to population growth. *The Economist* also published a number of articles outlining the debates and challenges of population growth in October 2011. On environmental Malthusianism, Frank Furedi discusses the limitations of Malthusian approaches in *Population and Development*, published by Polity Press, 1997. This can be read against classic texts on the limits of population growth, such as Ehrlich's *Population Bomb* and the Club of Rome's *Limits to Growth*.

On population ageing, the UNFPA's review of World Population Ageing 1950–2050 is available at www.un.org/esa/population/publications/worldageing19502050/. The UN report on replacement migration is also available online: www.un.org/esa/population/publications/migration/migration.htm

# REFERENCES

Aalbers, M. (2005) 'Place based social exclusion: redlining in the Netherlands', *Area* 37 (1): 100–9.

Amato, P.R. (2000) 'The consequences of divorce for adults and children', *Journal of Marriage and the Family*, 62 (4): 1269–87.

Amnesty International/Simon Russell (1999) *Most Vulnerable of all: The Treatment of Unaccompanied Refugee Children in the UK*. London: Amnesty International.

Anderson, G. and Hussey, P. (2000) 'Population aging: a comparison among industrialized countries', *Health Affairs*, 19 (3): 191–203.

Anderson, M. (1995) *Approaches to the History of theWestern Family, 1500–1914*. Cambridge: Cambridge University Press.

Babb, P., Martin, J. and Haezewindt, P. (eds) (2004) *Focus on Social Inequalities*. London: HMSO.

Bailey, A. (2009) 'Population geography: lifecourse matters', *Progress in Human Geography*, 33 (3): 407–18.

Bailey, A. and Boyle, P. (2004) 'Untying and retying family migration in the New Europe', *Journal of Ethnic and Migration Studies*, 30 (2): 229–41.

Bajekal, M. (2000) 'National health surveys', in D. Leadbetter (ed.) *Harnessing Official Statistics*. Abingdon: Radcliffe Medical Press, pp. 93–106.

Bajekal, M. (2005) 'Healthy life expectation by area deprivation: magnitude and trends in England, 1994–1999', *Health Statistics Quarterly*, 25.

Bajekal, M., Harries, T., Breman, R. and Woodfield, K. (2003) *Review of Disability Estimates and Definitions*. London: HMSO.

Bajekal, M. and Prescott, A. (2003) *Disability. Health Survey for England 2001*. London: The Stationery Office.

Bartley, M. (2004) *Health Inequality: An Introduction to Theories, Concepts and Methods*. Cambridge: Polity Press.

Bauman, Z. (2001) *The Individualised Society*. Cambridge: Polity Press.

Bauman, Z. (2002) 'Foreword', in U. Beck and E. Beck-Gernsheim (eds), *Individualisation: Institutionalised Individualism and its Social and Political Consequences*. London: Sage.

BBC (2004) *If … The Generations Fall Out*. Retrieved September 2011 from http://news.bbc.co.uk/1/hi/programmes/if/3489560.stm

BBC (2005) Mandela's eldest son dies of AIDS. Retrieved December 2011 from http://news.bbc.co.uk/1/hi/world/africa/4151159.stm

Beaujeu-Garnier, J. (1978) *Geography of Population*, 2nd edn. London: Longman.

Beck, U. and Beck-Gernsheim, E. (2002) *Individualisation: Institutionalised Individualism and its Social and Political Consequences*. London: Sage.

Becker, G.S. (1981) *A Treatise on The Family*. Cambridge, MA: Harvard University Press.

Beck-Gernsheim, E. (2002) *Reinventing the Family*. Cambridge: Polity Press.

Bederman, G. (2008) 'Sex, scandal, satire, and population in 1798: revisiting Malthus's first essay', *Journal of British Studies*, 47 (4): 768–95.

Bell, M., Blake, M., Boyle, P., Duke-Williams, O., Rees, R. and Stillwell, J. (2002) 'Cross national comparison of internal migration: issues and measures', *Journal of the Royal Statistical Society: Series A*, 165 (3): 435–64.

Bellaby, P. (2006) 'Can they carry on working? Later retirement, health and social inequality in an aging population', *International Journal of Health Services*, 36 (1): 1–23.

Benzeval, M., Judge, K. and Whitehead, M. (1995) 'Introduction', in M. Benzeval, K. Judge, and M. Whitehead (eds), *Tackling Inequalities in Health*. London: Kings Fund Publishing, pp, 1–9.

Berney, L., Blane, D., Davey Smith, G. and Holland, P. (2000) 'Lifecourse influences on health in early old age', in H. Graham (ed.), *Understanding Health Inequalities*. Buckingham: Open University Press, pp. 79–95.

Bernhardt, E. (1993) 'Fertility and employment', *European Sociological Review*, 9 (1): 25–42.

Best, R. (1999) 'Health inequalities: the place of housing. Inequalities in health' in Gordon, D., Shaw, M., Dorling, D. and Davey Smith, G. (eds) *The Evidence Presented to the Independent Inquiry into Inequalities in Health, Chaired by Sir Donald Acheson*. Bristol: The Policy Press, pp 45–67.

Billari, F.C., Philipov, D. and Baizán Muñoz, P. (2001) 'Leaving home in Europe: the experience of cohorts born around 1960', *International Journal of Population Geography*, 7 (5): 311–38.

Black, D., Morris, J., Townsend, P., Davidson, N. and Whitehead, M. (1982) *Inequalities in Health: The Black Report and The Health Divide*. London: Penguin.

Blaxter, M. (1990) *Health and Lifestyles*. London: Routledge.

Bond, J. and Rodriguez Cabrero, G. (2007) 'Health and dependency in later life', in J. Bond, S. Peace, F. Dittmann-Kohli and G. Westerhof, G. (eds), *Ageing in Society*. London: Sage, pp. 113–41.

Bongaarts, J. (1978) 'A framework for analysing the proximate determinants of fertility', *Population and Development Review*, 4 (1), 105–32.

Bongaarts, J. (2002) 'The end of fertility transition in the developed world', *Population and Development Review*, 28 (3), 419–43.

Bongaarts, J. (2003) 'Completing the fertility transition in the developing world: the role of educational differences and fertility preferences', *Population Studies*, 57 (3): 321–36.

Bongaarts, J. (2006) 'The causes of stalling fertility transitions', *Studies in Family Planning*, 37 (1): 1–16.

Bongaarts, J. and Bulatao, R.A. (2000) *Panel on Population Projections, Committee on Population National Research Council*. Washington, DC: National Academy Press.

Bos, D. and Weizsacker, R. (1989) 'Economic consequences of an aging population', *European Economic Review*, 33: 345–54.

Boseley, S. (2000) 'Mbeki insists poverty causes AIDS', *Guardian*. Retrieved on 7 March 2011 from www.guardian.co.uk/world/2000/jul/10/sarahboseley1

Bowling, A. (2005) 'Just one question: If one question works, why ask several?', *Quality and Safety in Health Care*, 59(5): 342–45.

Boyle, P., Halfacree, K. and Robinson, V. (1998) *Exploring Contemporary Migration*. Harlow: Longman.

Brettell, C. and Hollifield, J.F. (2008) (eds) *Migration Theory. Talking Across Disciplines*, 2nd edn. London: Routledge.

Brewster, K.L. and Rindfuss, R.R. (2000) 'Fertility and women's employment in industrialised nations', *Annual Review of Sociology*, 26 (1): 271–96.

Brimblecombe, N., Dorling, D. and Shaw, M. (1999) 'Mortality and migration in Britain, first results from the British Household Panel Survey', *Social Science and Medicine*, 49 (7): 981–8.

Brown, J.C. and Guinnane, T.W.G. (2002) 'Fertility transition in a rural, Catholic population: Bavaria 1880–1910', *Population Studies*, 56 (1): 35–49.

Brown, L.R. (1988) 'The changing world food prospect: the nineties and beyond', *Worldwatch Paper* 85, Washington, DC: Worldwatch Institute.

Budak, M.-A. E., Liaw, K.-L. and Kawabe, H. (1996) 'Co-residence of household heads with parents in Japan: a multivariate explanation' *International Journal of Population Geography*, 2 (1): 133–52.

Burrell, K. (2009) (ed.) *Polish Migration to the UK in the New European Union: After 2004.* Farnham: Ashgate.

Bushin, N. (2009) 'Researching family migration decision-making: a children-in-families approach', *Population, Space and Place*, 15 (5): 429–43.

Buzar, S., Ogden P.E. and Hall, R. (2005) 'Households matter: the quiet demography of urban transformation', *Progress in Human Geography*, 29 (4): 413–36.

Caldwell, J. (1982) *Theory of Fertility Decline.* New York: Academic Press.

Caldwell, J. (1986) 'The routes to low mortality in poor counties', *Population and Development Review*, 12 (2): 171–220

Caldwell, J. (1997) 'The global fertility transition: the need for a unifying theory', *Population and Development Review*, 23 (4): 803–12.

Caldwell, J.C. and Caldwell, P. (1987) 'The cultural context of high fertility in sub-Saharan Africa', *Population and Development Review*, 13 (3): 409–37.

Caldwell, J., Caldwell, P. and Orubuloye, P. (1992) 'Fertility decline in Africa: a new type of transition?', *Population and Development Review*, 18(2): 211–43.

Canning, D. (2011) 'The causes and consequences of demographic transition', *Population Studies*, 65 (3): 353–61.

Castles, F. (2003) 'The world turned upside down: below replacement fertility, changing preferences and family-friendly public policy in 21 OECD countries', *Journal of European Social Policy*, 13 (3): 209–27.

Castles, S. and Miller, D. (2009) *The Age of Migration*, 4th edn. Basingstoke: Palgrave.

Champion A. (1989) *Counterurbanisation: The Changing Pace and Nature of Population Deconcentration.* London: Edward Arnold.

Champion, A.G. and Fielding, A. (1993) (eds) *Migration Processes and Patterns: Research Progress and Prospects.* Chichester: John Wiley and Sons.

Chan, T.W. and Halpin, B. (2003) 'Union dissolution in the United Kingdom', *International Journal of Sociology*, 32 (4): 76–93.

Chandler, J., Williams, M., Maconachie, M., Collett, T. and Dodgeon, B. (2004) 'Living alone: its place in household formation and change', *Sociological Research Online*, 9 (3). Retrieved from www. socresonline. org.uk/9/3/chandler.html.

Chandola, T. and Jenkinson, C. (2000) 'Validating self-rated health in different ethnic groups', *Ethnicity and Health*, 5 (2): 151–9.

Chant, S. (1997) *Women-headed Households: Diversity and Dynamics in the Developing World.* Basingstoke: Macmillan.

Chant, S. (2004) 'Dangerous equations? How female-headed households became the poorest of the poor: causes, consequences and cautions', *IDS Bulletin*, 35 (4): 19–26.

Charles, E.N., Cheesman, T. and Hoffmann, S. (eds) (2003) *Between a Mountain and a Sea. Refugees writing in Wales.* Swansea: Hafan Books.

Charlton, J. (2000) 'ONS data: other health sources', in D. Leadbetter (ed.), *Harnessing Official Statistics.* Oxford: Radcliffe Medical Press, pp. 35–50.

Cherlin, A. (1999) 'Going to extremes: family structure, children's wellbeing and social science', *Demography*, 36 (4), 421–8.

Chesnais, J. (1996) 'Fertility, family and social policy in contemporary Western Europe', *Population and Development Review*, 22 (4): 729–39

Clarke, J.I. (1972) *Population Geography*, 2nd edn. Oxford: Pergamon Press.

Cleland, J. (2001) 'The effects of improved survival on fertility: a reassessment', *Population Development Review*, 27 (1): 60–92.

Cleland, J. and Wilson, C. (1987) 'Demand theories of the fertility transition: an iconoclastic view', *Population Studies*, 41 (1): 5–30.

Coale, A.J. and Cotts Watkins, S. (eds) (1986) *The Decline of Fertility in Europe*. Princeton, NJ: Princeton University Press.

Cohen, R. (2006) *Migration and Its Enemies*. Farnham: Ashgate.

Cole, K. (1994) 'Data modification, data suppression, small populations and other features of the 1991 small area statistics' *Area*, 26 (1): 69–78.

Connolly, M. (2008) *Fatal Misconception: The Struggle to Control the World Population*. Cambridge, MA: Harvard University Press.

Cooke, T., Boyle, P., Couch, K. and Feijten, P. (2009) 'A longitudinal analysis of family migration and the gender gap in earnings in the United States and Great Britain', *Demography*, 46 (1): 147–67.

Council of Europe (2004) *Demographic Year Book*. Strasbourg: Council of Europe Publishing.

Dale, A., Fieldhouse, E. and Holdsworth, C. (2000) *Analysing Census Microdata*. London: Arnold.

Dale, A. and Marsh, C. (eds) (1993) *The 1991 Census User's Guide*. London: HMSO.

Davey Smith, G., Dorling, D., Mitchell, R. and Shaw, M. (2002) 'Health inequalities in Britain: continuing increases up to the end of the 20th century', *Journal of Epidemiological Community Health*, 56: 434–5.

Davey Smith, G., Shaw, M. and Dorling, D. (1998) 'Shrinking areas and mortality', *The Lancet*, 352 (9138): 1439–40.

De Haas, H. (2010) 'Migration transitions: a theoretical and empirical inquiry into the developmental drivers of international migration', International Migration Institute, Oxford, Working Paper 24.

Department for Communities and Local Government (no date) Household Projections. Table 404: Household Projections 1 by Household Type and Region, England, 2001–33. Retrieved November 2011 from www.communities.gov.uk/housing/housingresearch/housingstatistics/housingstatisticsby/householdestimates/livetables-households/

Diamond, I. and Jefferies, J. (2001) *Beginning Statistics: An Introduction for Social Scientists*. London: Sage.

Dorling, D. (2000) 'The ghost of Christmas past: health effects of poverty in London in 1896 and 1991', *British Medical Journal*, 321 (7276): 1547–51.

Dorling, D. (2010) *Injustice: Why Social Inequality Persists*. Bristol: Policy Press.

Dorling, D., Shaw, M. and Brimblecombe, N. (2000) 'Housing wealth and community wealth: exploring the role of migration', in H. Graham (ed.), *Understanding Health Inequalities*. Buckingham: Open University Press, pp. 186–202.

Dorling, D. and Thomas, B. (2004) *People and Places: A 2001 Census Atlas of the UK*. Bristol: Policy Press.

Drinkwater, S., Eade, J. and Garapich, M. (2009) 'Poles apart? EU Enlargement and the labour market outcomes of immigrants in the United Kingdom', *International Migration*, 47 (1): 161–90.

Dubuc, S. (2009) 'Application of the own-children method for estimating fertility by ethnic and religious groups in the UK', *Journal of Population Research*, 26: 207–25.

Durand, J. and Massey, D.S. (2004) *Crossing the Border: Research from the Mexican Migration Project*. New York: Russell Sage Foundation.

Dyson, T. (2010) *Population and Development: The Demographic Transition*. London: Zed Books.

Dyson, T. (2011) 'The role of the demographic transition in the process of urbanization', *Population and Development Review,* 37: 34–54.

Ehrlich, P.R. (1968) *The Population Bomb.* New York: Ballantine Books.

Elliott, J. and Richards, M. (1991) 'Children and divorce: educational performance and behaviour before and after parental separation', *International Journal of Law and the Family,* 5: 258–76.

Ermisch, J.F. and Francesconi, M. (2000) 'The increasing complexity of family relationships: lifetime experience of lone motherhood and stepfamilies in Great Britain', *European Journal of Population,* 16: 235–49.

Ermisch, J.F. and Overton, E. (1985) 'Minimal household units: a new approach to the analysis of household formation', *Population Studies,* 39: 33–54.

Esping-Andersen, G. (1999) *Social Foundations of Post-industrial Economies.* Oxford: Oxford University Press.

Eurostat (2009) *Youth in Europe a Statistical Portrait.* Luxembourg: Publications Office of the European Union.

Eurostat Data Service, European Commission (no date) Main Tables. Retrieved December 2010 from http://epp.eurostat.ec.europa.eu/portal/page/portal/population/data/main_tables

Evans, R., McGrail, K., Morgan, S., Barer, M. and Hertzman, S. (2001) 'Apocalypse no: population aging and the future of health care systems', *Canadian Journal on Aging,* 20 (1): 160–91.

Fields, J. (2003) 'America's families and living arrangements 2003', *Current Population Reports, P20–553.* United States Census Bureau, Washington, DC.

Findlay, A. (1988) 'From settlers to skilled transients: the changing structure of British international migration', *Geoforum,* 19: 401–10.

Finney, N. and Simpson, L. (2008) 'Internal migration and ethnic groups: evidence for Britain from the 2001 Census', *Population, Space and Place,* 14: 63–83.

Finney, N. and Simpson, L. (2009) '*Sleepwalking to Segregation'? Challenging Myths of Race and Migration.* Bristol: Policy Press.

Flowerdew, R. and Al-Hamad, A. (2004) 'The relationship between marriage, divorce and migration in a British dataset', *Journal of Ethnic and Migration Studies',* 30: 339–51.

Folbre, N. (1991) 'Women on their own: residential independence in Massachusetts in 1880', *Continuity and Change,* 6 (1): 87–105.

Ford, J., Rudd, J. and Burrows, R. (2002) 'Conceptualising the contemporary role of housing in the transition to adult life in England', *Urban Studies December,* 39: 2455–67.

Fotheringham, A.S., Brunsdon, C. and Charlton, M. (2005) *Quantitative Geography: Perspectives on Spatial Data Analysis.* London: Sage.

Freer, C. (1988) 'Old myths: frequent misconceptions about the elderly', in N. Wells and C. Freer (eds), *The Ageing Population: Burden or Challenge.* London: Macmillan, pp. 3–16.

Frejka, T. and Sobotka, T. (2008) 'Overview Chapter 1: fertility in Europe: diverse, delayed and below replacement', *Demographic Research,* 19 (3): 15–46. Retrieved from www.demographic-research.org/volumes/vol19/3/

Frey, W. (1995) 'Immigration and internal migration flight from US metropolitan areas: towards a new demographic Balkanisation', *Urban Studies,* 32(4–5): 733–57.

Fuller, E. (2010) 'Adult alcohol consumption. health Survey for England – 2009', In R. Craig and V. Hirani (eds), *Health and Lifestyles.* Vol. 1, Chapter 10, pp. 167–86. The Health and Social Care Information Centre, Leeds.

Furedi, F. (1997) *Population and Development: A Critical Introduction.* Cambridge: Polity Press.

Gatrell, A.C. and Elliott, S.J. (2009) *Geographies of Health: An Introduction.* Oxford: Wiley-Blackwell.

Giddens, A. (1992) *The Transformation of Intimacy: Sexuality, Love and Eroticism in Modern Societies.* Cambridge: Polity Press.

Glick-Schiller, N. and Faist, T. (2010) (eds) *Migration, Development and Transnationalization: A Critical Stance.* Oxford: Berghahn Books.

Goldscheider, F. and Goldscheider, C. (1999) *The Changing Transition to Adulthood.* London: Sage.

Gordon, M. (1964) *Assimilation in American Life: The Role of Race, Religion and National Origins.* Oxford: Oxford University Press.

Gorman, M. (2011) *Life Expectancy: Myth or Reality.* Retrieved October 2012 from http://www.helpage.org/blogs/mark-gorman-25/life-expectancy-myth-and-reality-374/

Gould, M. and Jones, K. (1996) 'Analyzing perceived limiting long-term illness using U.K. census microdata', *Social Science and Medicine*, 42 (6): 857–69.

Gould, W.T.S. (2009) *Population and Development.* London: Routledge.

Government Actuary Department (2006) *Fertility Assumptions.* Retrieved October 2011 from www.gad.gov.uk/Demography%20Data/Population/2006/methodology/fertass.html

Graham, E., Macleod, M., Johnston, M., Dibben, C., Morgan, I. and Briscoe, S. (2000) 'Individual deprivation, neighbourhood and recovery from illness', in H. Graham (ed.), *Understanding Health Inequalities.* Buckingham: Open University Press, pp. 170–85.

Graham, H. (2000) 'The challenge of health inequalities', in H. Graham (ed.), *Understanding Health Inequalities.* Buckingham: Open University Press.

Greenwood, S. (2008) *Just One Child: Science and Policy in Deng's China.* Berkeley: University of California Press.

Guy, W. (1996) 'Health for all', in R. Levitas, and W. Guy (eds), *Interpreting Official Statistics.* London: Routledge.

Hajnal, J. (1965) 'European marriage patterns in perspective', in D.V. Glass and D.E.C. Everseley (eds), *Population in History.* London, Arnold.

Halfacree, K. (2008) 'To revitalise counterurbanisation research? Recognising an international and fuller picture', *Population, Space and Place*, 14: 479–95.

Halfacree, K. and Boyle, P. (1993) 'The challenge facing migration research: the case for a biographical approach', *Progress in Human Geography*, 17: 333–58.

Hall, R. and Ogden, P.E. (2003) 'The rise of living alone in Inner London: trends among the population of working age', *Environment and Planning A*, 35: 871–88.

Hall, R., Ogden, P.E. and Hill, C. (1999) 'Living alone: evidence from England and Wales and France for the last two decades', in S. McRae (ed.), *Changing Britain: Families and Households in the 1990s.* Oxford: Oxford University Press.

Halliday, S. (2000) 'William Farr: campaigning statistician', *Journal of Medical Biography*, 8 (4): 220–7.

Hanson, J. and Bell, M. (2007) 'Harvest trails in Australia: patterns of seasonal migration in the fruit and vegetable industry', *Rural Studies*, 23: 101–17.

Haskey, J. (1996) 'Families and household in Great Britain', *Population Trends*, 85: 7–24.

Haskey, J. (2005) 'Living arrangements in contemporary Britain: having a partner who usually lives elsewhere and living apart together (LAT)', *Population Trends*, 122: 35–45.

Hay, S., Cox, G.D., Rogers, D., Randolph, S., Sternk, D., Shanks, D., Myers, M. and Snow, R. (2002) 'Climate change and the resurgence of malaria in the East African highlands', *Nature*, 415 (21 Feb): 905–9.

Heath, S. and Cleaver, E. (2003) *Young, Free and Single. Twenty-Somethings and Household Change.* Basingstoke: Palgrave Macmillan.

Henshall Momsen, J. (2002) 'Myth or math: the waxing and waning of the female-headed household', *Progress in Development Studies*, 2 (2): 145–51.

Higgins, V., Afkhami, R., Meadows, G. and Rafferty, A. (2007) *Introductory Guide to Using the Large-scale Government Surveys for Health Research. ESDS Data Guides*. Manchester: University of Manchester.

Hinde, A. (1998) *Demographic Methods*. Arnold: London.

Hobcraft, J. (1996) 'Fertility in England and Wales: a fifty year perspetive', *Population Studies*, 50 (3): 485–524.

Hogan, D.P. and Goldsheider, F.K. (2003) 'Success and challenge in demographic studies of the life course' in Jeylan T. Mortimer and Michael J. Shanahan (eds), *Handbook of the Life Course*. New York: Kluwer Academic/Plenum Publishers, pp. 681–91.

Holdsworth, C. (2000) 'Leaving home in Great Britain and Spain', *European Sociological Review*, 16 (2): 201–22.

Holdsworth, C. (2004) 'Family support and the transition out of parental home in Britain, Spain and Norway', *Sociology*, 38 (5): 909–26.

Home Office (2006) *Control of Immigration Statistics 2005*. London: Home Office.

Hunt, S. (2005) *The Life Course. A Sociological Introduction*. Basingstoke: Palgrave Macmillan.

Hvistendahl, M. (2011) 'Young and restless can be a volatile mix', *Science*, 333 (29 July): 552–4.

Iacovou, M. (2002) 'Regional differences in the transition to adulthood', *Annals of the American Academy of Political and Social Science*, 580 (1): 40–69.

Iacavou, M. and Berthoud, R. (2001) *Young People's Lives: A Map of Europe*. Institute for Social and Economic Research: University of Essex.

Idler, E. and Benyamini, I. (1997) 'Self rated health and mortality: a review of 27 community studies', *Journal of Health and Social Behaviour*, 38 (1): 21–37.

International Organisation for Migration (no date) Facts and Figures. Retrieved January 2010 from www.iom.int/jahia/Jahia/about-migration/facts-and-figures/lang/en

Jagger, C., Cox, B. and Le Roy, S. (2006) 'Healthy expectancy calculation by the Suillivan method', EHEMU Technical Report, European Health Expectancy Monitoring Unit.

Japan Statistics Bureau (2006) *Census of Japan 2005. Sex, Age and Marital Status of Population, Structure and Housing Conditions of Households*. Retrieved December 2010 from www.stat.go.jp/english/data/kokusei/index.htm

Japan Statistics Bureau (2011) *Statistical Handbook of Japan 2010*. Retrieved December 2010 from www.stat.go.jp/english/data/handbook/index.htm

Japan Statistics Bureau (no date) *Historical Data: Private Households, Household Members and Related Members 65 Years of Age and Over (Private Households with Related Members 75 Years of Age and Over, and 85 Years of Age and Over) by Family Type of Household 1980–2005*. Retrieved from www.stat.go.jp/english/data/chouki/02.htm

Jarvis, C. (2000) 'Trends in old age morbidity and disability in Britain', *Ageing and Society*, 19: 603.

Jones, G. (1995) *Leaving Home*. Buckingham: Open University Press.

Joshi, H. (2002) 'Production, reproduction, and education: women, children, and work in a British perspective', *Population and Development Review*, 28: 445–74.

Joshi, H., Wiggins, R., Bartley, M., Mitchell, R., Gleave, S. and Lynch, K. (2000) 'Putting health inequalities on the map: does where you live matter, and why?', in H. Graham (ed.) *Understanding Health Inequalities*. Buckingham: Open University Press, pp. 143–55.

Jouvenal, H. (1989) *Europe's Ageing Population: Trends and Challenges to 2025*. Guildford: Butterworth and Co.

Kalmijn, M. (2007) 'Explaining cross-national differences in marriage, cohabitation, and divorce in Europe, 1990–2000', *Population Studies*, 61 (3): 243–63.

Kalra, V. and Kapoor, N. (2009) 'Interrogating segregation, integration and the community cohesion agenda', *Journal of Ethnic and Migration Studies*, 35 (9).

Kamerman, S.B. and Kahn, A.J. (1988) *Mothers Alone: Strategies for a Time of Change*. Dover, MA: Auburn House Publishing Co.

Kertzer, D. (1997) 'The proper role of culture in demographic explanation.' in G. Jones, R. Douglas, J. Caldwell, and R. D'Souza (eds), *The Continuing Demographic Transition*. Oxford: Clarendon Press, pp. 137–57.

Kiernan, K. (1996) 'Cohabitation in Western Europe', *Population Trends*, Summer, 25–32.

Kiernan, K. (2004) 'Unmarried cohabitation and parenthood in Britain and Europe', *Law and Policy*, 26 (1): 33–55.

Kiernan, K. and Mueller, G. (1999) 'Who divorces?' in S. McRae (ed.), *Changing Britain: Families and Households in the 1990s*. Oxford: Oxford University Press, pp. 377–403.

Kirk, D. (1996) 'Demographic transition theory', *Population Studies*, 50 (3): 361–87.

Kohler, H-P., Billari, F.C and Ortega, J.A. (2006) 'Low fertility in Europe: causes, implications and policy options'. Retrieved from www.ssc.upenn.edu/~hpkohler/papers/Low-fertility-in Europe-final.pdf

Koser, K. (2007) *International Migration. A Very Short Introduction*. Oxford: Oxford University Press.

Kreider, R.M. and Elliott, D.B. (2009) 'America's families and living arrangements: 2007', *Current Population Reports, P20–561*. Washington, DC: United States Census Bureau.

Kreider, R.M. and Fields, J.M. (2001) 'Number, timing, and duration of marriages and divorces: Fall 1996', *Current Population Reports, P70–80*. Washington, DC: United States Census Bureau.

Kujisten, A.C. (1996) 'Changing family patterns in Europe: a case study of divergence?', *European Journal of Population*, 12: 115–43.

Kulu, H. and Billari, F.C. (2004) 'Multilevel analysis of internal migration in a transitional country: the case of Estonia', *Regional Studies*, 38: 679–96.

Kulu, H. and Boyle, P. (2010) 'Premarital cohabitation and divorce: support for the "Trial Marriage" Theory?'. *Demographic Research*, 23 (31). Retrieved from www.demographic-research.org/volumes/vol23/31/

Kunst, A. E. (2005) 'Trends in socioeconomic inequalities in self-assessed health in 10 European countries', *International Journal of Epidemiology* 34 (2): 295–305.

Kunzig, K. (2011) 'Population 7 bullion', *National Geographic*, January 2011.

Laslett, T.P.R. (1965) 'Misbeliefs about our ancestors' in Laslett, T.P.R., *The World We Have Lost*. London: Routledge.

Laughlin, L. (2010) 'Who's minding the kids? Child care arrangements: Spring 2005/Summer 2006', *Current Population Reports, P70–121*. Washington, DC: United States Census Bureau.

Lee, R. (2011) 'The outlook for population growth', *Science*, 333 (29 July), 569–73.

Lesthaege, R. (1995) 'The second demographic transition in western countries: an interpretation', in K. Oppenheim Mason and A. Jensen (eds), *Gender and Family Change in Industrialized Countries*. Oxford: Clarendon Press.

Lesthaeghe, R. and Neels, K. (2002) 'From the first to the second demographic transition: an interpretation of the spatial continuity of demographic innovation in France, Belgium and Switzerland', *European Journal of Population*, 18 (4): 325–60.

Levin, I. (2004) 'Living apart together: a new family form', *Current Sociology*, 52 (2): 223–40.

Lindley, Anna and Van Hear, N. (2007) *New Europeans on the Move: A preliminary review of the onward migration of refugees within the European Union*. Working Paper. No. 57. Centre on Migration, Policy and Society, University of Oxford.

Livi-Bacci, M. (2001) *A Concise History of World Population*, 3rd edition. Oxford: Blackwell.

Lloyd, C.D. (2010) *Spatial Data Analysis: An Introduction for GIS Users*. Oxford: Oxford University Press.

Lloyd, C.D. (2011) *Local Models for Spatial Analysis*, 2nd edition. Boca Raton: CRC Press.

Lomborg, B. (2001) *The Skeptical Environmentalist: Measuring the Real State of the World*. Cambridge: Cambridge University Press.

Macintyre, S., Hiscock, R., Kearns, A. and Ellaway, A. (2000) 'Housing tenure and health inequalities: a three dimensional perspective on people, homes and neighbourhoods', in H. Graham (ed.), *Understanding Health Inequalities*. Buckingham: Open University Press, pp. 129–42.

Malakoff, D. (2011) 'Are more people necessarily a problem?', *Science* 333 (29 July): 544–6.

Malthus, T. (1970) *An Essay on the Principle of Population and A Summary View of the Principle of Population*, edited by A. Flew. Harmondsworth: Penguin.

Manor, O., Matthews, S. and Power, C. (2001) 'Self-rated health and limiting longstanding illness: inter-relationships with morbidity in early adulthood', *International Journal of Epidemiology*, 30 (3): 600–7.

Marmot, M., Feeney, A., Shipley, M., North, F. and Syme, S. (1995) 'Sickness absence as a measure of health status and functioning: from the UK Whitehall II study', *Journal of Epidemiology and Community Health*, 49 (2): 124–30.

Marsh, C. (1993) 'An overview', in Angela Dale and Cathie Marsh (eds), *The 1991 Census User's Guide* London: HMSO.

Marsh, C. and Elliott, J. (2008) *Exploring Data*. Cambridge: Polity Press.

Marshall, A. (2009) 'Developing a methodology for the estimation and projection of limiting long term illness and disability', PhD thesis, School of Social Sciences, University of Manchester.

Martin, D. (1995) 'Censuses and the modelling of population in GIS', in Paul Longley and Graham Clarke (eds), *GIS for Business and Service Planning*. Cambridge: GeoInformation International.

Mason, J. (2004) 'Personal narratives, relational selves: residential histories in the living and telling', *The Sociological Review*, 52 (2): 162–79.

Mason, K.O. (1997) 'Explaining fertility transitions', *Demography* 34 (4): 443–54.

Mathews, T.J., Hamilton, B.E. (2002) 'Mean age of mother, 1970–2000', *National Vital Statistics Reports*, 51 (1). Hyattsville, MD: National Centre for Health Statistics.

Matthews, R., Jagger, C. and Hancock, R. (2006) 'Does socio-economic advantage lead to a longer healthier life', *Social Science and Medicine*, 62 (10): 2489–99.

Mberu, B.U. (2007) 'Household structure and living conditions in Nigeria', *Journal of Marriage and Family*, 69: 513–27.

McCracken, K. and Phillips, D. (2005) 'International demographic transitions', in G. Andrews and D. Phillips (eds), *Ageing and Place: Perspectives, Policy and Practice*. Abingdon: Routledge, pp 36–60.

McCrone, P., Dhanasiri, S., Patel, A., Knapp, M. and Lawton-Smith, S. (2008) *Paying the Price: The Cost of Mental Health in England to 2026*. London: London School of Economics.

McIssac Cooper, S. (1999) 'Historical analysis of the family', in Marvin B. Sussman, Suzanne K. Steinmetz, and Gary W. Peterson (eds) *Handbook of Marriage and the Family*. New York: Plenum Press, pp. 13–38.

McKeown, T. (1976) *The Modern Rise of Population*. London: Arnold.

McRae, S. (1999) 'Introduction', in S. McRae (ed.), *Changing Britain: Families and Households in the 1990s*. Oxford: Oxford University Press.

Meadows, D.H. (1974) *Limits to Growth: A Report for the Club of Rome's Project on the Predicament of Mankind*. London: Pan.

Merriman, P. and Cresswell, T. (2008) *Mobilities: Practices Spaces Subjects*. Farnham: Ashgate.

Mitchell, R. (2005) 'Commentary: the decline of death–how do we measure and interpret changes in self-reported health across cultures and time?', *International Journal of Epidemiology*, 34 (2): 306–8.

Moon, G. and Gould, M. (2000) *Epidemiology: An Introduction*. Buckingham: Open University Press.

Mullan, P. (2000) *The Imaginary Timebomb*. London: I.B. Tauris.

National Statistics Office, Republic of the Philippines (2003) 'Number of live births by age-specific fertility Rate and total fertility rate: Philippines'. Retrieved October 2011 from www.census.gov.ph/data/sectordata/2003/sr0620703.htm

Newell, C. (1988) *Methods and Models in Demography*. Chichester: Wiley.

Ní Bhrolcháin, M. (2001) 'Divorce effects and causality in the social sciences', *European Sociological Review (Special Issue on Causality)*, 17 (1): 33–57.

Ní Bhrolcháin, M., Chappell, R., Diamond, I, and Jameson, C. (2000) 'Parental divorce and outcomes for children', *European Sociological Review*, 16 (1): 67–92.

Ní Laoire, C. (2000) 'Conceptualising Irish rural youth migration: a biographical approach', *International Journal of Population Geography*, 6, 229–43.

Norman, P. (1997) 'Small area population updates', Research Section, City of Bradford Metropolitan District Council.

Norman, P. and Bambra, C. (2007) 'Incapacity or unemployment? The utility of an administrative data source as an updatable indicator of population health', *Population, Space and Place*, 13 (5): 333–52.

Norman, P., Boyle, P. and Rees, P. (2005) 'Selective migration health and deprivation: a longitudinal analysis', *Social Science and Medicine*, 60 (12): 2755–71.

Norman. P., Rees, P., Wohland, P. and Boden, P. (2010) 'Ethnic group populations: the components for projection, demographic rates and trends', in J. Stillwell and M. van Ham (eds) *Ethnicity and Integration*, Understanding Population Trends and Processes Series. Dordrecht: Springer, 289–315.

OECD (2007) Social Expenditure database 1980–2003. Retrieved December 2009 from www.oecd.org/els/social/expenditure

OECD (2011) Family Database, OECD, Paris. Retrieved December 2011 from www.oecd.org/social/family/database

ONS (2000) *Social Trends 30*. London: The Stationery Office.

ONS (2003) *Households*. Retrieved December 2010 from www.statistics.gov.uk/cci/nugget.asp?id=350

ONS (2004) *Methods for National Statistics 2001 Area Classification for Local Authorities*. Retrieved from www.statistics.gov.uk/about/methodology_by_theme/area_classification/la/methodology.asp

ONS (2007) *Focus on Families*. London: The Stationery Office.

ONS (no date) Discussion Paper: Population Base for 2011 Census Enumeration.

Openshaw, S. (ed.) (1995) *Census Users' Handbook*. Cambridge: GeoInformation International.

Openshaw, S. and Wymer, C. (1995) 'Classifying and regionalising census data', in Stan Openshaw (ed.) in *Census Users' Handbook*. Cambridge: GeoInformation International.

O'Reilly, D., Rosato, M. and Patterson, C. (2005) 'Self reported health and mortality: ecological analysis based on electoral wards across the United Kingdom', *British Medical Journal*, 331 (7522): 938–9.

Parkinson, M., Champion, T., Evans, R., Simmie, J., Turok, I., Cookston, M., Katz, B., Park, A., Berube, A., Coombes, M.G., Dorling. D., Glass, N., Hutchins, M., Kearns, A., Martin, R., Wood, P. (2006) *State of the English Cities.* London: Office of the Deputy Prime Minister.

Peach, C. (1996a) 'Does Britain have ghettos?', *Transactions of the Institute of British Geographers*, 21 (1): 216–35.

Peach, C. (1996b) 'Good segregation, bad segregation', *Planning Perspectives*, 11: 1–20.

Peach, C. (2009) 'Slippery segregation: discovering or manufacturing ghettos?', *Journal of Ethnic and Migration Studies* 35 (9) 1381–95.

Pearce, D. (1999) 'Changes in fertility and family sizes in Europe', *Population Trends*, Spring, 33–40.

Phillimore, P. and Morris, D. (1991) 'Discrepant legacies: premature mortality in two industrial towns', *Social Science and Medicine,* 33 (2): 139–52.

Pooley, C. and Turnbull, J. (2004) 'Migration from the parental home in Britain since the eighteenth century', in F. van Poppel, M. Oris and J. Lee (eds), *The Road to Independence*. Oxford: Peter Lang, pp. 375–402.

Portes, A. and DeWind, J. (2008) *Rethinking Migration: New Theoretical and Empirical Perspectives.* Oxford: Berghahn Books.

Portes, A. and Zhou, M. (1993) 'The new second generation: segmented assimilation and its variants', *The Annals of the American Association of Political and Social Science*, 530 (1): 74–96.

Preston, S. (2007) 'The changing relation between mortality and level of economic development', *International Journal of Epidemiology*, 36: 484–90.

Price, S. (2000) 'Hospital episode statistics', in D. Leadbetter (ed.), *Harnessing Official Statistics.* Abingdon: Radcliffe Medical Press, pp. 51–62.

Ravenstein, E.G. (1885) 'The laws of migration', *Journal of the Statistical Society of London*, 48 (2): 167–235.

Ravenstein, E.G. (1889) 'The laws of migration', *Journal of the Royal Statistical Society*, 52 (2): 241–305.

Raymer, J. and Willekens, F. (2008) (eds) *International Migration in Europe: Data, Models and Estimates*. Chichester: John Wiley and Sons.

Rees, P. (1996) 'Access to population census data for research purposes in the UK', paper presented at the Annual Conference of the Australian Population Association, University of Adelaide, 5/12/96.

Rees, P. (2008) What happens when international migrants settle? Projections of ethnic groups in United Kingdom regions, in J. Raymer and F. Willekens, F. (ed.), *International Migration in Europe: Data, Models and Estimates*. Chichester: John Wiley and Sons, pp. 329–58.

Rees, P., Parsons, J. and Norman, P. (2005) 'Making an estimate of the number of people and households for Output Areas in the 2001 Census', *Population Trends*, 122: 27–34.

Rees, P., Wohland, P., Norman, P. and Boden, P. (2011) 'A local analysis of ethnic group population trends and projections for the UK', *Journal of Population Research*, 28 (2): 129–48, DOI: 10.1007/s12546-011-9047-4.

Reher, D. (2004) 'The demographic transition revisited as a global process', *Population Space and Place* 10: 19–41.

Riley, Jason L. (2008) *Let Them In: The Case for Open Border*. New York: Gotham.

Rogerson, P.A. (2006) *Statistical Methods for Geography*. London: Sage.

Rowland, D.T. (2003) *Demographic Methods and Concepts*. Oxford: Oxford University Press.

Ruggles, S. (2009) 'Reconsidering the North-West European family system', *Population and Development Review*, 35 (2): 249–73.

Sachs, J. (2007) Bursting at the seams. Reith lecture. Retrieved December 2011 from www.bbc.co.uk/print/radio4/reith2007/lecture1.shtml?print

Samers, M. (2010) *Key Ideas in Geography: Migration*. London: Routledge.

Savage, M. and Burrows, R. (20090 'Some further reflections on the coming crisis of empirical sociology', *Sociology*, 43 (4): 765–75.

Scherbov, S. and Sanderson, W. (2011) 'Rethinking ageing', British Society for Population Studies.

Senior, M. and Viveash, B. (1998) *Health and Illness*. London: Palgrave Macmillan.

Shaw, M., Dorling, D., Gordan, D. and Davey Smith, G. (1999) *The Widening Gap: Health Inequalities and Policy in Britain*. Bristol: The Policy Press.

Shaw, M., Dorling, D. and Mitchell, R. (2002) *Health, Place, and Society*. Harlow: Pearson.

Shkolnikov, V., McKee, M., Leon, D.A. (2001) 'Changes in life expectancy in Russia in the mid-1990s', *The Lancet*, 357 (9260): 917–21.

Simmons, T. and O'Neill, G. (2001) *Households and Family 2000: Census 2000 Brief C2KBR/01-8*. U.S. Washington, DC: US Census Bureau.

Simon, J. L. (1982) *The Ultimate Resource*. Princeton, NJ: Princeton University Press.

Simpson, L. and Finney, N. (2009) 'Spatial patterns of internal migration: evidence for ethnic groups in Britain', *Population, Space and Place*, 15, 37–56.

Skeldon, R. (1997) *Migration and Development: A Global Perspective*. Harlow: Longman.

Smart, C. (2007) *Personal Life*. Cambridge: Polity Press.

Smart, C. and Neale, B. (1999) *Family Fragments*. Cambridge: Polity Press.

Smith, K., Downes, B. and O'Connell, M. (2001) 'Maternity leave and employment patterns: 1961–1995', *Current Population Reports*, P70–79. Washington, DC: United States Census Bureau.

Snowdon, P. (2010) *The Spirit Level Delusion: Fact-checking the Left's New Theory for Everything*. London: Little Dice.

Solinger, D. (1999) 'Citizenship issues in China's internal migration: comparisons with Germany and Japan', *Political Science Quarterly*, 114 (3): 455–78.

Staines, A. (1999) 'Poverty and health', in D. Dorling and S. Simpson (eds), *Statistics in Society: The Arithmetic of Politics*. London: Arnold, pp. 252–62.

Statistics Sweden (no date) *Statistical Database*. Retrieved December 2010 from www.ssd.scb.se/databaser/makro/start.asp?lang=2

Stillwell, J., Rees, P. and Boden, P (eds) (1991) *Migration Processes and Patterns: Population Redistribution in the UK*. Chichester: John Wiley and Sons.

Thane, P. (1989) 'Old age: burden or benefit?', in H. Joshi (ed.), *The Changing Population of Britain*. Oxford: Blackwell, pp. 46–55.

The Club of Rome (no date) 'The birth of the Club of Rome'. Retrieved October 2011 from www.clubofrome.org/?p=375

Tiffen, M., Mortimore, M. and Gichuki F. (1994) *More People Less Erosion: Environmental Recovery in Kenya*. Chichester: John Wiley and Sons.

UNAIDS (2010) UN AIDS Report on the Global AIDS Epidemic. Joint United Nations Programme on HIV/AIDS (UNAIDS) Geneva: UN. Retrieved December 2011 from www.unaids.org/globalreport/Global_report.htm.

UN Children's Fund (2002) *The State of the World's Children 2002: Leadership*. Geneva: UNICEF.

UN Children's Fund (2007) *The State of the World's Children 2008: Child Survival* Geneva: UNICEF.

UN Department of Economic and Social Affairs Population Division (2000) *Replacement Migration: Is It a Solution to Declining and Ageing Populations?* New York: UN. Retrieved December 2011 from www.un.org/esa/population/publications/migration/migration.htm

UN Department of Economic and Social Affairs Population Division (2004) *World Population Monitoring 2002 Reproductive Rights And Reproductive Health*. New York: UN.

UN Department of Economic and Social Affairs Population Division (2006) *Trends in Total Migration Stock: The 2005 Revision*. New York: UN. Retrieved from www.un.org/esa/population/publications/migration/UN_Migrant_Stock_Documentation_2005.pdf

UN Department of Economic and Social Affairs, Population Division (2009a) *Trends in International Migrant Stock: The 2008 Revision*. New York: UN. Retrieved April 2011 from www.un.org/esa/population/migration/UN_MigStock_2008.pdf

UN Department of Economic and Social Affairs, Population Division (2009b) *World Population Prospects. The 2008 Revision*. New York: UN. Retrieved from www.un.org/esa/population/publications/wpp2008/wpp2008_highlights.pdf

UN Department of Economic and Social Affairs Population Division (2011) *World Population Projections 2010 Revision*. New York: UN. Retrieved December 2011 from http://esa.un.org/unpd/wpp/unpp/panel_population.htm

UN Department of Economic and Social Affairs Population Division (no date) *Assumptions Underlying the 2010 Revision*. Retrieved October from http://esa.un.org/unpd/wpp/documentation/WPP2010_ASSUMPTIONS_AND_VARIANTS.pdf

UN Department of Economic and Social Affairs, Statistics Division (1998) *Principles and Recommendations for National Population Censuses,* Series M, 67. New York: UN.

UN Department of Economic and Social Affairs, Statistics Division (2005a) *Demographic Yearbook 2004*. New York: UN. Retrieved June 2010 from http://unstats.un.org/unsd/demographic/products/dyb/dyb2.htm

UN Department of Economic and Social Affairs, Statistics Division (2005b) *Special Report of the World's Women 2005: Progress in Statistics. Focusing on Sex-disaggregated Statistics on Population, Births and Deaths*. New York: UN.

UN Department of Economic and Social Affairs, Statistics Division (2006) *Demographic Yearbook 2005*. New York: UN. Retrieved June 2010 from http://unstats.un.org/unsd/demographic/products/dyb/dyb2.htm

UN Department of Economic and Social Affairs, Statistics Division (2008) *Gender Info 2007*. New York: UN. Retrieved from http://unstats.un.org/unsd/demographic/products/genderinfo/

UN Department of Economic and Social Affairs, Statistics Division (2011) 2010 *World Population and Housing Census Programme*. New York: UN. Retrieved September 2011 from http://unstats.un.org/unsd/demographic/sources/census/2010_PHC/default.htm

UN Human Settlements Programme (2001) *Cities in A Globalizing World – Global Report on Human Settlements*. London: Earthscan.

Urry, J. (2007) *Mobilities*. Cambridge; Polity Press.

US Census Bureau (2004a) 'Table HH-4. Households by size: 1960 to present'. Retrieved from http://www.census.gov/population/socdemo/hh-fam/tabHH-4.pdf

US Census Bureau (2004b) *Table AD-1. Young Adults Living At Home: 1960 to Present*. Retrieved December 2010 from www.census.gov/population/socdemo/hh-fam/tabAD-1.pdf

US Census Bureau (2006) 'Table MS-2. Estimated median age at first marriage, by sex: 1890 to the present'. Retrieved December 2010 from www.census.gov/population/socdemo/hh-fam/ms2.pdf

US Census Bureau (no date) 'Families and living arrangements'. Retrieved December 2011 from www.census.gov/population/www/socdemo/hh-fam.html

Van de Kaa, D.J. (1987) 'The second demographic transition', *Population Bulletin* 42 (1). Washington, DC: The Population Reference Bureau.

Van de Kaa, D.J. (1996) 'Anchored narratives: the story and findings of half a century of research into the determinants of fertility', *Population Studies*, 50 (3): 389–432.

Van de Kaa, D.J. (1999) 'Without maps and compass? Towards a new European transition project', *European Journal of Population*, 15 (4): 309–16.

Verdon, M. (1998) *Rethinking Households: An Atomistic Perspective on European Living Arrangements*. London: Routledge.

Vertovec, S. (2003) 'Migration and other modes of transnationalism: towards conceptual cross-fertilization', *International Migration Review*, 37 (3): 641–65.

Vickers, D.W. and Rees, P.H. (2006) 'Introducing the national classification of census output areas', *Population Trends*, 125: 15–29.

Wall, R. (1978) 'The age at leaving home', *Journal of Family History*, 3: 181–202.

Wardle, A. (2010) 'Cigarette smoking. Health survey for England 2009', in R. Craig and V. Hirani (eds), *Health and Lifestyles*, Vol. 1, pp. 151–66. Health and Social Care Information Centre, Leeds.

Westerlund, H., Kiyimaki, M., Singh-Manoux, A., Melcior, M., Ferrie, J., Pentti, J., Jokela, J., Leineweber, C., Goldberg, M., Zins, M. and Vahtera, J. (2009) 'Self-rated health before and after retirement in France (GAZEL): a cohort study', *The Lancet*, 374.

Whitehead, M., Diderichson, F. and Burstrom, B. (2000) 'Researching the impact of public policy on inequalities in health', in H. Graham (ed.), *Understanding Health Inequalities*. Buckingham: Open University Press, pp. 203–18.

WHO (2008a) *The Global Burden of Disease: 2004 Update*. Geneva: World Health Organization.

WHO (2008b) *Closing the Gap in a Generation: Health Equity through Action on the Social Determinants of Health. Final Report of the Commission on Social Determinant of Health*. Geneva: World Health Organization.

WHO (2011) 'Life tables for member states', Retrieved December 2011 from www.who.int/healthinfo/statistics/mortality_life_tables/en/

Wilkinson, R. (1997) 'Socioeconomic determinants of health: Health inequalities: relative or absolute material standards?', *British Medical Journal*, 314 (7080): 591–5.

Wilkinson, R. and Pickett, K. (2009) *The Spirit Level: Why Greater Equality Makes Societies Stronger*. New York: Bloomsbury Press.

Williams, G., Meth, P. and Willis, K. (2009) *Geographies of Developing Areas: The Global South in an Changing World*. Oxford: Routledge.

Winch, D. (1987) *Malthus*. Oxford: Oxford University Press.

Wingens, M., Windzio, M., de Valk, H. and Aybek, C. (2011) (eds) *A Life-Course Perspective on Migration and Integration*. Dordrecht: Springer.

Woods, R.I. (2006) *Children Remembered: Responses to Untimely Death in the Past*. Liverpool: Liverpool University Press.

World Bank (2006) *World Development Indicators*. Washington, DC: World Bank.

World Bank (2010) 'Outlook for Remittance Flows 2010–11', *Migration and Development Brief* 12 (April 23). Retrieved October 2011 from http://go.worldbank.org/SSW3DDNLQ0

Yeoh, B. (2007) 'Singapore: hungry for foreign workers at all skill levels', *Migration Information Source*. Retrieved December 2011 from http://www.migrationinformation.org/feature/display.cfm?ID=570

Zelinsky, W. (1971) 'The hypothesis of the mobility transition', *The Geographical Review*, 61 (2): 219–49.

Zweifel, P., Felder, S. and Meirs, M. (1999) 'Ageing of the population and health care expenditure: a red herring?' *Health Economics*, 8 (6): 485–96.

# INDEX

'bt', 'f' and 'b' following page numbers refer to information in tables, figures and boxes respectively.